Multiple Constant Multiplication Optimizations for Field Programmable Gate Arrays

Martin Kumm

Multiple Constant Multiplication Optimizations for Field Programmable Gate Arrays

With a preface by Prof. Dr.-Ing. Peter Zipf

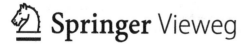 Springer Vieweg

Martin Kumm
Kassel, Germany

Dissertation of Martin Kumm in the Department of Electrical Engineering and Computer Science at the University of Kassel. Date of Disputation: October 30th, 2015

ISBN 978-3-658-13322-1   ISBN 978-3-658-13323-8 (eBook)
DOI 10.1007/978-3-658-13323-8

Library of Congress Control Number: 2016935387

Springer Vieweg
© Springer Fachmedien Wiesbaden 2016

Printed on acid-free paper

This Springer Vieweg imprint is published by Springer Nature
The registered company is Springer Fachmedien Wiesbaden GmbH

Preface

As silicon technology advances, field programmable gate arrays appear to gain ground against the traditional ASIC project starts, reaching out to form the mainstream implementation basis. Their predefined structures result in an essential inefficiency, or performance gap at all relevant axes, i.e. clock frequency, power and area. Thus, highly optimised system realisations become more and more important to use this technology at its best. Microarchitectures and their adaptation to the FPGA hardware, combined with an optimal matching of model structures and FPGA structures, are two points of action where engineers can try to get optimally balanced solutions for their designs, thus fighting the performance gap towards the ASIC reference.

While microarchitecture design based on the knowledge of FPGA structures is located in the domain of traditional hardware engineering, the mapping and matching is based on EDA algorithms and thus strongly related to computer science. Algorithms and the related sophisticated tools are permanently in short supply for leading edge optimisation needs.

Martin's dissertation deals with the algorithmic optimisation of circuits for the multiplication of a variable with constants in different flavours. As this type of operations is elementary in all areas of digital signal processing and also usually on the critical path, his approaches and results are of high relevance not only by themselves but also as a direction for further research. His direct contributions are the advancement of algorithmic treatment of pipeline-based multiplier circuits using heuristics and exact optimisation algorithms, the adaptation of several algorithms to the specific conditions of field programmable gate arrays, specifically lookup-table based multipliers, ternary adders and embedded multipliers with fixed word length, and the transfer of his findings to a floating-point multiplier architecture for multiple constants.

Along with all the accompanying details, this is a large range of topics Martin presents here. It is an impressive and comprehensive achievement, convincing by its depth of discussion as well as its contributions in each of the areas.

Martin was one of my first PhD candidates. Thrown into the cultural conflict between computer science and electrical engineering he soon developed a sense for the symbiosis of both disciplines. The approach and results are symptomatical for this developing interdisciplinary area, where systems and their optimisation algorithms are developed corporately.

I found Martin's text exciting to read, as it is a comprehensive work within the ongoing discussion of algorithmic hardware optimisation. His work is supported by a long list of successful publications. It is surely a contribution worth reading.

Kassel, 22. of April, 2016   *Prof. Dr.-Ing. Peter Zipf*

Abstract

Digital signal processing (DSP) plays a major role in nearly any modern electronic system used in mobile communication, automotive control units, biomedical applications and high energy physics, just to name a few. A very frequent but resource intensive operation in DSP related systems is the multiplication of a variable by several constants, commonly denoted as multiple constant multiplication (MCM). It is needed, e. g., in digital filters and discrete transforms. While high performance DSP systems were traditionally realized as application specific integrated circuits (ASICs), there is an ongoing and increasing trend to use generic programmable logic ICs like field programmable gate arrays (FPGAs). However, only little attention has been paid on the optimization of MCM circuits for FPGAs.

In this thesis, FPGA-specific optimizations of MCM circuits for low resource usage but high performance are considered. Due to the large FPGA-specific routing delays, one key for high performance is pipelining. The optimization of pipelined MCM circuits is considered in the first part of the thesis. First, a method that optimally pipelines a given (not necessary optimal) MCM circuit using integer linear programming (ILP) is proposed. Then, it is shown that the direct optimization of a pipelined MCM circuit, formally defined as the pipelined MCM (PMCM) problem, is beneficial. This is done by presenting novel heuristic and optimal ILP-based methods to solve the PMCM problem. Besides this, extensions to the MCM problem with adder depth (AD) constraints and an extension for optimizing a related problem – the pipelined multiplication of a constant matrix with a vector – are given.

While these methods are independent of the target device and reduce the number of arithmetic and storage components, i. e., adders, subtracters and pipeline registers, FPGA-specific optimizations are considered in the second part of the thesis. These optimizations comprise the inclusion of look-up table (LUT)-based constant multipliers, embedded multipliers and the use of ternary (3-input) adders which can be efficiently mapped to modern FPGAs. In addition, an efficient architecture to perform MCM operations in floating point arithmetic is given.

Abstract

Acknowledgements

First of all, I would like to thank my supervisor, Prof. Dr.-Ing. Peter Zipf, who gave me the opportunity to do my PhD under his supervision. He provided an excellent guidance and was always open-minded to new and exceptional ideas, leading to creative and fruitful discussions. I also thank my colleagues at the digital technology group of the University of Kassel. In special, Michael Kunz, who had many ideas how to prepare the course material and Konrad Möller, who started as student with his project work and master's thesis and is now a highly-valued colleague. Our mathematicians, Diana Fanghänel, Dörthe Janssen and Evelyn Lerche, were always open to help with math related questions. The deep understanding in discrete optimization and ILP (and more) of Diana Fanghänel helped a lot in obtaining some of the results in this work. I am grateful to all the students that contributed to parts of this thesis with their bachelor or master's thesis, project work or student job: Martin Hardieck, who did a great job in the C++ implementation of RPAG extensions for ternary adders and the constant matrix multiplication (and the combination of both) during his project work, his diploma thesis and his student job; Katharina Liebisch, for implementing MCM floating point units for reference; Jens Willkomm, who realized the ternary adder at primitive level for Xilinx FPGAs during his project work; and Marco Kleinlein, who implemented flexible VHDL code generators based on the Floating Point Core Generator (FloPoCo) library [1]. I would also like to thank our external PhD candidates Eugen Bayer and Charles-Frederic Müller for interesting discussions and collaborative work. I also appreciate the financial support of the German research council which allowed me to stay in a post-doc position in Kassel.

During the PhD, I started cooperations with Chip Hong Chang and Mathias Faust from the Nanyang Technological University, Singapore, Oscar Gustafsson and Mario Garrido from the University of Linköping, Sweden, and Uwe Meyer-Baese from the Florida State University, USA. Especially, I am very grateful to Oscar Gustafsson to invite me as guest researcher in his group. Exchanging ideas with people from the same field expanded my horizon enormously. Thanks to the ERASMUS program, two visits in

Linköping were possible and further visits are planned. Another big help was the friendly support of other researchers. Levent Aksoy was always fast in answering questions, providing filter benchmark data or even the source code of their algorithms. My thanks also go to Florent de Dinechin and all the contributors of the FloPoCo library [1] for providing their implementations as open-source [2]. Open-source in research simplifies the comparison with other work which inspired me to publish the presented algorithms as open-source, too.

Last but not least, I want to kindly thank my wife Nadine and my son Jonathan. Their love and support gave me the energy to create this thesis.

Contents

9 Floating Point Multiple Constant Multiplication   **153**

9.1   Constant Multiplication using Floating Point Arithmetic . . . 153

9.2   Floating Point MCM Architecture 154

    9.2.1   Multiplier Block 154

    9.2.2   Exponent Computation 155

    9.2.3   Post-Processing 156

9.3   Experimental Results 156

    9.3.1   Floating Point SCM 157

    9.3.2   Floating Point MCM 158

9.4   Conclusion 161

10 Optimization of Adder Graphs with Ternary (3-Input) Adders   163

10.1   Ternary Adders on Modern FPGAs 163

    10.1.1   Realization on Altera FPGAs 164

    10.1.2   Realization on Xilinx FPGAs 165

    10.1.3   Adder Comparison 166

10.2   Pipelined MCM with Ternary Adders 167

    10.2.1 $\mathcal{A}$-Operation and Adder Depth 168

    10.2.2   Optimization Heuristic RPAGT 168

    10.2.3   Predecessor Topologies 169

10.3   Experimental Results 172

10.4   Conclusion 174

11 Conclusion and Future Work   **177**

11.1   Conclusion 177

11.2   Future Work 178

    11.2.1   Use of Compressor Trees 178

    11.2.2   Reconfigurable MCM 179

    11.2.3   Optimal Ternary SCM circuits 179

List of Figures

List of Tables

Abbreviations

AD	adder depth
ALM	adaptive logic module
ALUT	adaptive LUT
ASAP	as soon as possible
ASIC	application specific integrated circuit
BDFS	bounded depth-first search (algorithm)
BHM	Bull and Horrocks modified (algorithm)
BILP	binary integer linear programming
BLE	basic logic element
CFGLUT	configurable LUT
CLB	complex logic block
CMM	constant matrix multiplication
CPA	carry propagate adder
CSA	carry-save adder
CSD	canonic signed digit
CSE	common subexpression elimination
CSR	cut-set retiming
DA	distributed arithmetic
DAG	directed acyclic graph
DCT	discrete cosine transform
DFS	depth-first search
DFT	discrete Fourier transform
DiffAG	difference adder graph (algorithm)

DSP	digital signal processing
FA	full adder
FF	flip flop
FFT	fast Fourier transform
FIR	finite impulse response
FloPoCo	Floating Point Core Generator
FPGA	field programmable gate array
FPMCM	floating point MCM
FPSCM	floating point SCM
GA	genetic algorithm
GPC	glitch path count
GPS	glitch path score
HA	half adder
HDL	hardware description language
HL	high level
IC	integrated circuit
IIR	infinite impulse response
ILP	integer linear programming
IP	intellectual property
ISE	Integrated Software Environment
LAB	logic array block
LE	logic element
LL	low level
LP	linear programming
LSB	least significant bit
LUT	look-up table
MAG	minimized adder graph (algorithm)
MCM	multiple constant multiplication
MILP	mixed integer linear programming

MSB	most significant bit
MSD	minimum signed digit
MST	minimum spanning tree
MTZ	Miller-Tucker-Zemlin
MUX	multiplexer
NOF	non-output fundamental
NRE	non-recurring engineering
OPAG	optimal pipelined adder graph (algorithm)
PAG	pipelined adder graph
PALG	pipelined adder/LUT graph
PCMM	pipelined CMM
PMCM	pipelined MCM
PSOP	pipelined SOP
RAM	random access memory
RAG-n	n-dimensional reduced adder graph (algorithm)
RCA	ripple carry adder
RMCM	reconfigurable MCM
RPAG	reduced pipelined adder graph (algorithm)
RPAGT	reduced pipelined adder graphs with ternary adders (algorithm)
RSG	reduced slice graph (algorithm)
SCM	single constant multiplication
SCP	set cover with pairs
SD	signed digit
SOP	sum-of-products
STL	standard template library
TO	timeout
ulp	unit in the last place
VHDL	very high speed integrated circuit hardware description language
VLSI	very large scale integrated

XDL   Xilinx design language

XOR   exclusive-or

XST   Xilinx Synthesis Technology

Author's Publications

The following publications were published during the time of the PhD preparation between May 2009 and August 2015. Most of them are included in this thesis.

[1] M. Kumm, H. Klingbeil, and P. Zipf, "An FPGA-Based Linear All-Digital Phase-Locked Loop," *IEEE Transactions on Circuits and Systems I: Regular Papers*, vol. 57, no. 9, pp. 2487–2497, 2010.

[2] M. Kumm and P. Zipf, "High Speed Low Complexity FPGA-Based FIR Filters Using Pipelined Adder Graphs," in *IEEE International Conference on Field-Programmable Technology (FPT)*, 2011, pp. 1–4.

[3] M. Kumm, P. Zipf, M. Faust, and C.-H. Chang, "Pipelined Adder Graph Optimization for High Speed Multiple Constant Multiplication," in *IEEE International Symposium on Circuits and Systems (ISCAS)*, 2012, pp. 49–52.

[4] M. Kumm, K. Liebisch, and P. Zipf, "Reduced Complexity Single and Multiple Constant Multiplication in Floating Point Precision," in *IEEE International Conference on Field Programmable Logic and Application (FPL)*, 2012, pp. 255–261.

[5] M. Kunz, M. Kumm, M. Heide, and P. Zipf, "Area Estimation of Look-Up Table Based Fixed-Point Computations on the Example of a Real-Time High Dynamic Range Imaging System," in *IEEE International Conference on Field Programmable Logic and Application (FPL)*, 2012, pp. 591–594.

[6] M. Kumm and P. Zipf, "Hybrid Multiple Constant Multiplication for FPGAs," in *IEEE International Conference on Electronics, Circuits and Systems, (ICECS)*, 2012, pp. 556–559.

[7] U. Meyer-Baese, G. Botella, D. Romero, and M. Kumm, "Optimization of High Speed Pipelining in FPGA-Based FIR Filter Design Using Genetic Algorithm," in *Proceedings of SPIE*, 2012, pp. 1–12.

[8] M. Kumm, D. Fanghänel, K. Möller, P. Zipf, and U. Meyer-Baese, "FIR Filter Optimization for Video Processing on FPGAs," *EURASIP Journal on Advances in Signal Processing (Springer)*, pp. 1–18, 2013.

[9] M. Kumm, K. Möller, and P. Zipf, "Reconfigurable FIR Filter Using Distributed Arithmetic on FPGAs," in *IEEE International Symposium on Circuits and Systems (ISCAS)*, 2013, pp. 2058–2061.

[10] M. Kumm, K. Möller, and P. Zipf, "Partial LUT Size Analysis in Distributed Arithmetic FIR Filters on FPGAs," in *IEEE International Symposium on Circuits and Systems (ISCAS)*, 2013, pp. 2054–2057.

[11] M. Kumm, M. Hardieck, J. Willkomm, P. Zipf, and U. Meyer-Baese, "Multiple Constant Multiplication with Ternary Adders," in *IEEE International Conference on Field Programmable Logic and Application (FPL)*, 2013, pp. 1–8.

[12] M. Kumm, K. Möller, and P. Zipf, "Dynamically Reconfigurable FIR Filter Architectures with Fast Reconfiguration," *International Workshop on Reconfigurable Communication-centric Systems-on-Chip (ReCoSoC)*, 2013, pp. 1–8.

[13] M. Kumm and P. Zipf, "*Efficient High Speed Compression Trees on Xilinx FPGAs*," in *Methoden und Beschreibungssprachen zur Modellierung und Verifikation von Schaltungen und Systemen (MBMV)*, 2014, pp. 171–182.

[14] K. Möller, M. Kumm, B. Barschtipan, and P. Zipf, "Dynamically Reconfigurable Constant Multiplication on FPGAs." in *Methoden und Beschreibungssprachen zur Modellierung und Verifikation von Schaltungen und Systemen (MBMV)*, 2014, pp. 159–169.

[15] M. Kumm and P. Zipf, "Pipelined Compressor Tree Optimization Using Integer Linear Programming," in *IEEE International Conference on Field Programmable Logic and Applications (FPL)*, 2014, pp. 1–8.

[16] K. Möller, M. Kumm, M. Kleinlein, and P. Zipf, "Pipelined reconfigurable multiplication with constants on FPGAs," in *IEEE International Conference on Field Programmable Logic and Application (FPL)*, 2014, pp. 1–6.

[17] M. Kumm, S. Abbas, and P. Zipf, "An Efficient Softcore Multiplier Architecture for Xilinx FPGAs," in *IEEE Symposium on Computer Arithmetic (ARITH)*, 2015, pp. 18–25.

[18] M. Faust, M. Kumm, C.-H. Chang, and P. Zipf, "Efficient Structural Adder Pipelining in Transposed Form FIR Filters," in *IEEE International Conference on Digital Signal Processing (DSP)*, 2015, pp. 311-314.

[19] K. Möller, M. Kumm, C.-F. Müller, and P. Zipf, "Model-based Hardware Design for FPGAs using Folding Transformations based on Subcircuits," in *International Workshop on FPGAs for Software Programmers (FSP)*, 2015, pp. 7–12

[20] M. Garrido, P. Källström, M. Kumm, and O. Gustafson, "CORDIC II: A New Improved CORDIC Algorithm," *IEEE Transactions on Circuits and Systems II: Express Briefs*, accepted for publication in 2015.

1 Introduction

The multiplication of a variable by several constants, commonly denoted as multiple constant multiplication (MCM), is a frequent operation in many DSP systems. It occurs in digital filters as well as discrete transforms like, e.g., the fast Fourier transform (FFT) or the discrete cosine transform (DCT). It can be found in a variety of signal processing areas such as communication, feedback control, image processing, automotive applications, biomedical applications, radar and high energy physic's experiments [10–13]. However, the MCM operation is resource intensive compared to other common signal processing operations like additions or delays. Hence, a rich body of literature exists dealing with its complexity reduction. Most of these techniques reduce the constant multiplications to additions, subtractions and bit-shift operations. The conventional approach is to minimize the number of add/subtract operations or their underlying building blocks like full adders (FAs) and half adders (HAs). Their target are typically very large scale integrated (VLSI) circuits, realized as application specific integrated circuits (ASICs). This work considers the optimization of the MCM operation for FPGAs. Why the FPGA-specific optimization is an important goal is motivated in the following.

1.1 Motivation

While performance demanding systems were built using ASICs in the past, they are nowadays more and more often replaced by FPGAs. There are several reasons for this, one reason is the flexibility of FPGAs. Their functionality is (re-)programmable leading to reduced development time, lower non-recurring engineering (NRE) cost and, thus, a shorter time-to-market. Another reason are the increasing ASIC costs. One large part in the ASIC manufacturing costs is the mask costs, which roughly double with each process generation [14]. This leads to prohibitive manufacturing prices. About four million dollars have to be raised for a full mask set at the 40 nm technology node [15]. FPGAs are manufactured in high-volume production which allows the latest cutting-edge technology. This allows designs with very low quantities to benefit from high-end technologies.

However, FPGAs consist of completely different resources compared to the gate-level cells of standard cell libraries. The typical FPGA resources are look-up tables (LUTs), fast carry chains, block RAMs and embedded multipliers which are organized in a fixed layout. In addition, the routing delays between these resources are much more dominant compared to the wire delays in VLSI designs. Therefore, a good conventional MCM solution may perform poor when mapped to an FPGA. The importance of the MCM operation and the particularities of FPGAs necessitate the development of specific optimization methods which are considered in this work.

1.2 Goal and Contributions of the Thesis

The main purpose of this thesis is the investigation of optimization methods and hardware structures to efficiently map MCM operations to FPGAs. An efficient mapping is regarded as a realization of a circuit which uses minimal FPGA resources and power consumption and a maximal throughput. Besides the optimization of MCM operations, there are two closely related problems. First, the single constant multiplication (SCM) corresponds to the scaling operation which is a special case of MCM. Second, the MCM operation can be generalized to several input variables leading to the multiplication of a constant matrix with a vector, denoted as constant matrix multiplication (CMM) in the following. This work focusses on MCM but implications to these related operations are discussed wherever applicable. All common FPGA resources are considered including fast carry chains, look-up tables (LUTs) and flip flop (FF) as well as embedded multipliers.

The major contributions of this thesis are as follows:

- An ILP-based optimization method for optimally pipelining a predefined adder graph

- A heuristic, called reduced pipelined adder graph (RPAG), for the direct optimization of pipelined adder graphs (PAGs) (the pipelined MCM (PMCM) problem) which is alternatively capable to solve the MCM problem with constrained adder depth (AD)

- An optimal ILP-based method for the PMCM problem which is also suitable for MCM with AD constraints

- An extension of the RPAG heuristic to optimize circuits for the multiplication of a constant matrix with a vector (pipelined or with AD constraints)

- An extension of the optimal PMCM method to incorporate LUT-based multipliers

- An extension of the RPAG heuristic to incorporate embedded multipliers/DSP blocks

- An efficient architecture to compute MCM in floating point arithmetic

- An extension of the RPAG heuristic for the efficient use of ternary adders

- Open-source implementations of most of the proposed algorithms [16]

1.3 Organization of the Thesis

This thesis is divided into two parts. The first part considers the pipelined MCM (PMCM) problem and the second part considers FPGA-specific MCM optimizations. In the remaining section of this chapter, the most important MCM applications are further detailed to motivate this work. The background of the two main parts is provided in Chapter 2. First, the problem of constructing shift-and-add-based constant multiplications is introduced starting with SCM which is then extended to MCM and CMM. Next, the representation of the shift-and-add-based MCM operation as a so-called adder graph, some properties and common terms as well as the related optimization problems are introduced in Section 2.4. The architecture of FPGAs and the relevant aspects for this thesis are introduced in Section 2.5. The related work is given in Section 2.6. Chapter 2 serves as a background for the following chapters. After reading the background chapter, each chapter of the two main parts can be read independently. For the experienced reader, Chapter 2 may be skipped and may serve as a reference for definitions used in later chapters.

One key observation from the FPGA considerations in Section 2.5 is that the routing delay in a shift-and-add-based MCM operation has a large impact on the total delay, resulting in a worse performance for FPGAs. The counter-measure to this performance degradation is pipelining. In part one (chapters 3-6), pipelining shift-and-add-based MCM operations is considered. The relation between the chapters of the two main parts is visualized

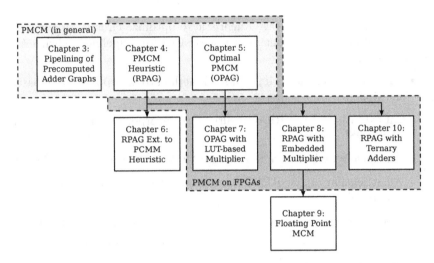

Figure 1.1: Graphical representation of the relation of the main thesis chapters 3-10

in Figure 1.1. Applying pipelining to a shift-and-add-based MCM circuit is not unique as there is a degree of freedom in which pipeline stage the additions or subtractions are scheduled. The problem of optimizing the pipeline for a given (precomputed) adder graph in a resource-optimal way is considered in Chapter 3. As this optimization strongly depends on the given adder graph, the MCM optimization problem is extended to the PMCM problem and a heuristic optimization method, called RPAG algorithm, is presented in Chapter 4. Optimal methods for solving the PMCM problem using different ILP formulations are given in Chapter 5. While chapters 3-5 provide solutions to the PMCM problem, part one is concluded with a (pipelined) CMM extension of the RPAG algorithm in Chapter 6.

In part two (chapters 7-10), MCM operations that include FPGA-specific resources like LUTs and embedded multipliers are considered. In Chapter 7, LUT-based constant multipliers are considered as an alternative for coefficients which are costly to realize in shift-and-add-based PMCM circuits. For that, the ILP formulation of Chapter 3 is extended. The inclusion of embedded multipliers in PMCM circuits is considered in Chapter 8. Here, the focus is on constants with large word size which typically appear in floating point applications. Therefore, the RPAG heuristic of Chapter 4 was extended to this kind of problem. An architecture for the MCM operation in floating point arithmetic is presented in Chapter 9. Finally, the RPAG algorithm is extended in Chapter 10 to the usage of ternary adders (i. e., three-input adders). Ternary adders can be mapped to the same resources as used by a common two-input adder on modern FPGAs of the two market leaders Xilinx and Altera.

1.4 Applications of Multiple Constant Multiplication

One of the most important building block in digital signal processing systems is the digital filter. It can be found in endless applications like in noise reduction, the suppression of unwanted signals, equalization of channels, sample rate conversion, etc. It is described by

$$y_n = \sum_{k=0}^{N-1} a_k x_{n-k} - \sum_{k=1}^{M-1} b_k y_{n-k} \qquad (1.1)$$

where a_k and b_k denote the filter coefficients and x_n and y_n denote the input and output samples of the filter, respectively [17]. The filter described by (1.1) may have an infinite impulse response (IIR), if $b_k = 0$ for $1 \leq k \leq M - 1$ it is a finite impulse response (FIR) filter[1]. Most MCM related publications are motivated by FIR filters as they have a linear phase but their required number of multiplications is typically much larger compared to an IIR filter with same magnitude response specification. Typically, the filter coefficients are obtained offline during the design time of the system and they are constants. The coefficient word sizes of practical FIR filters with approximation errors between -70 dB and -100 dB are between 15 bit and 20 bit [18].

The structure of an FIR filter in so-called direct form is shown in Figure 1.2(a). An alternative architecture can be obtained by transposing the structure of Figure 1.2(a). This is done by reversing the directions of each edge, replacing branches by adders and vice versa, and swapping the input and output [19]. The resulting structure is shown in Figure 1.2(b). In the direct form, several signals (the time shifted input signal) have to be multiplied by constants and added afterwards which is called a sum-of-products (SOP) operation (dashed box in Figure 1.2(a)). In the transposed form, a single

---

[1]Note that there exist filters with finite impulse response for which $b_k \neq 0$.

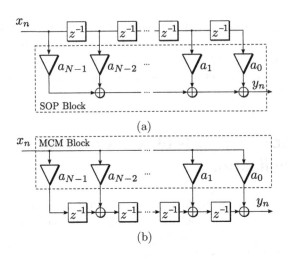

Figure 1.2: Structure of an FIR filter in (a) direct form and (b) transposed form

input has to be first multiplied by several constants which is called a multiplier block or MCM block (dashed box in Figure 1.2(b)). The remaining filter consists of adders and registers as shown in Figure 1.2(b). An N-tap FIR filter requires the multiplication with N constants. For linear phase filters which have symmetric coefficients, only about half of the multiplications are required. MCM and SOP operations are closely related as a shift-and-add-based MCM operation with a given set of constants can be converted to an equivalent SOP operation with the same constants by applying the transpose transformation and vice versa [20].

Another important application of MCM are linear discrete transforms expressed as

$$X_k = \sum_{k=0}^{N-1} x_k W_N^k \quad \text{for } k = 0 \dots N - 1 \tag{1.2}$$

where x_k are the input samples, X_k are the output samples and W_N^k is the filter kernel [17]. Different transformations can be computed by choosing the right filter kernel, e. g., choosing $W_N^k = e^{-j\frac{2\pi k}{N}}$ computes the discrete Fourier transform (DFT) while $W_N^k = \cos(\frac{2\pi k}{N})$ computes the DCT. One important application of these transformations is data compression as done in image (e. g., JPEG), video (e. g., MPEG) or audio (e. g., MP3) compression. The word lengths used in those applications are highly application dependent.

For the DFT (realized as FFT), word lengths between 5 bit [21] and 32 bit [22] were reported in literature.

Depending on the architecture, several constant multiplications occur. Take for example, a single complex multiplication that occurs in the FFT. It can be decomposed into two MCM instances (the real and imaginary part of the input) with two constants each (the real and imaginary part of the constant). Alternatively, it can be seen as a CMM instance of a two-dimensional vector with a 2×2 matrix. Other examples of linear transforms are the Walsh-Hadamard transform [23], kernels used in H.264/AVC [24] or color space conversions [25].

All these applications consist of SCM, MCM or CMM kernels which are – most of the time – the most resource demanding parts. In the next chapter, the background of add-and-shift-based constant multiplication schemes and their mapping to FPGAs is introduced.

2 Background

This chapter provides the background of the two main parts of this thesis. The circuit and graph representations of SCM, MCM and CMM operations are given and the notation and related work is introduced which is common for the later chapters.

2.1 Single Constant Multiplication

The multiplication with a single constant is introduced in the following, starting from the standard binary multiplication.

2.1.1 Binary Multiplication

A generic multiplication of two unsigned binary B-bit integer numbers $x = (x_{B-1} \ldots x_1 x_0)$ and $y = (y_{B-1} \ldots y_1 y_0)$, with $x_i, y_i \in \{0, 1\}$ can be written as

$$x \cdot y = x \cdot \left(\sum_{i=0}^{B-1} 2^i y_i \right) = \sum_{i=0}^{B-1} 2^i \underbrace{x \cdot y_i}_{\text{partial product}} \quad . \tag{2.1}$$

In a generic multiplier, the partial products $x \cdot y_i$ can be obtained by a bitwise AND-operation. The final product is then obtained by adding the bit-shifted partial products. Now, if y is a constant bit vector, all bits y_i which are zero lead to zero partial products and the corresponding adders can be removed. Thus, the number of required adders for the constant multiplication using this representation is equal to the number of non-zero elements (Hamming weight) in the binary representation of y minus one. To illustrate this, consider a multiplier with $y = 93$. Its binary representation $y = 1011101_2$ has five ones and requires four adders in the corresponding add-and-shift-realization $93x = (2^6 + 2^4 + 2^3 + 2^2 + 1)x$ as illustrated in Figure 2.1(a). Note that bit shifts to the left are indicated by left arrows in Figure 2.1.

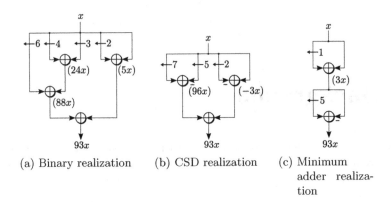

(a) Binary realization   (b) CSD realization   (c) Minimum
                                                     adder realiza-
                                                     tion

Figure 2.1: Different adder circuits to realize a multiplication with 93

2.1.2 Multiplication using Signed Digit Representation

The total number of operations can be further reduced by allowing subtract operations in the add-and-shift network. As the subtract operation is nearly equivalent in hardware cost (in terms of area, speed and energy), both are referred to as adders in the following. The adder reduction can be realized by converting the constant to the signed digit (SD) number system [26–28] (sometimes also called "ternary balanced notation" [29]), in which each digit is represented by one of the values of $\{-1, 0, 1\}$. In the example, the coefficient can be represented by $93 = 10\bar{1}00\bar{1}01_\mathrm{SD}$ where digit $\bar{1}$ corresponds to -1, i. e., $93 = 2^7 - 2^5 - 2^2 + 2^0$. Now, the corresponding circuit uses one adder less compared to the binary representation as illustrated in Figure 2.1(b).

The signed digit number system is not unique as there may exist several alternative representations for the same number. For example, 93 could also be represented by $93 = 1\bar{1}0\bar{1}000\bar{1}\,\bar{1}_\mathrm{SD}$, which has the same number of non-zero digits as the binary representation. A unique representation with minimal number of non-zero digits is given by the canonic signed digit (CSD) representation. A binary number can be converted to the CSD with the following simple algorithm [30, 31]:

Algorithm 1 (CSD Conversion Algorithm). *Starting with the least significant bit (LSB), search a bit string of the form '011...11' with length ≥ 2 and replace it with '10...0$\bar{1}$' of the same length. This procedure is repeated until no further bit string can be found.*

For example, the binary representation of $93 = 1011101_2$ is transformed to $93 = 1100\bar{1}01_{\mathrm{SD}}$ in the first iteration which is transformed to $93 = 10\bar{1}00\bar{1}01_{\mathrm{CSD}}$ in the second and final iteration. The CSD representation is unique and guarantees a minimal number of non-zeros. However, there may still be several SD representations with a minimal number of non-zeros. The representations with a minimal number of non-zeros are called minimum signed digit (MSD). The SD representation after the first iteration for 93 is an example of an MSD number which is not a CSD number.

Starting from the CSD representation, valid MSD representations can be constructed by replacing the bit patterns '$10\bar{1}$' with '011' or '$\bar{1}01$' with '$0\bar{1}\bar{1}$'. Doing this for all combinations results in a set of MSD numbers. For the example '93', four different MSD representations can be constructed:

$$\begin{aligned} 93 &= 10\bar{1}00\bar{1}01_{\mathrm{CSD}} \\ &= 01100\bar{1}01_{\mathrm{MSD}} \\ &= 10\bar{1}000\bar{1}\bar{1}_{\mathrm{MSD}} \\ &= 011000\bar{1}\bar{1}_{\mathrm{MSD}} \end{aligned} \tag{2.2}$$

2.1.3 Multiplication using Generic Adder Graphs

Although the arithmetic complexity of the constant multiplier can be reduced by using an MSD representation of the constant, it is not guaranteed to be minimal. Consider again the example number 93 which can be factored into $93 = 3 \cdot 31$. With $3 = 11_{\mathrm{MSD}}$ and $31 = 10000\bar{1}_{\mathrm{MSD}}$, the cascade of these two constant multipliers reduces the required adders to only two as shown in Figure 2.1(c).

This solution can be obtained by recognizing that two patterns in the CSD representation of 93 are related to each other:

$$10\bar{1}_{\mathrm{SD}} = -\bar{1}01_{\mathrm{SD}} \,. \tag{2.3}$$

With that, 93 can be represented as $93 = 10\bar{1}00\bar{1}01_{\mathrm{CSD}} = 3 \cdot 2^5 - 3 = 3 \cdot (2^5 - 1) = 3 \cdot 31$. This concept is known as sub-expression sharing and the corresponding optimization method is called common subexpression elimination (CSE) [12,32,33]. However, there is no guarantee to find a solution with the minimal number of adders as some common patterns may be hidden by other patterns due to overlaps [34]. Take for example, $25 = 10\bar{1}001_{\mathrm{CSD}}$ which can be factored into $25 = 5 \cdot 5 = 101_2 \cdot 101_2$. Due to the fact that two ones in the pattern 101_2 overlap when computing $25 = 101_2 + 2^2 \cdot 101_2 = 11001_2$, the corresponding 101_2 pattern is not visible in the CSD representation.

Finding an adder circuit that multiplies with a given single constant using a minimum number of adders is an optimization problem which is known as the SCM problem.

2.2 Multiple Constant Multiplication

A frequent application in signal processing is the multiplication of a variable by multiple constants which is commonly called multiple constant multiplication (MCM). Here, intermediate adder results can be shared between different coefficients such that the overall complexity is reduced. Take, for example, the multiplication with the constants 19 and 43. Their CSD based multipliers as introduced in the last section are shown in Figure 2.2(a). Redundancies may be found using CSE like in the SCM case but not only within a single coefficient but also between different coefficients. One MSD representation of the example constants is $19 = 10011_{MSD}$ and $43 = 101011_{MSD}$ (in which case the binary representations are MSD representations themselves). In both representations, the bit pattern 11_{MSD} is found which can be shared to reduce one adder. The resulting adder circuit is shown in Figure 2.2(b). However, as demonstrated in Figure 2.2(c) another adder configuration can be found which does the same multiplications with one adder less. It can not be obtained by the CSE approach as no bit pattern of the factor $5 = 101_{MSD}$ can be found in both MSD representations of the constants. The problem in finding the adder circuit with minimum adders is known as the MCM problem.

2.3 Constant Matrix Multiplication

An extension of the MCM operation (and the corresponding optimization problem) is the multiplication of a constant matrix with a vector of variables, which is called constant matrix multiplication (CMM) in the following. Its application can be found in digital filters (parallel 2D filters or polyphase filters [35, 36]), transformations like the FFT [37] or DCT [38], color space conversion in video processing [39] as well as in the multiplication of complex numbers by complex constants [40]. Again, the operation count can be reduced by sharing intermediate results. Take, for example, the complex

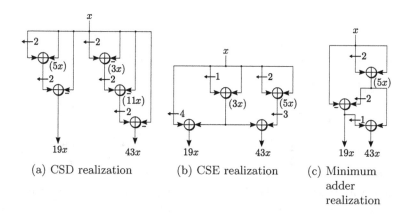

(a) CSD realization (b) CSE realization (c) Minimum adder realization

Figure 2.2: Different adder circuits to realize a multiplication with coefficients 19 and 43

multiplication

$$(x_r + jx_i) \cdot (19 + j43) = \underbrace{19x_r - 43x_i}_{=y_r} + j\underbrace{(43x_r + 19x_i)}_{=y_i} \qquad (2.4)$$

which can be represented in matrix form as

$$\begin{pmatrix} y_r \\ y_i \end{pmatrix} = \begin{pmatrix} 19 & -43 \\ 43 & 19 \end{pmatrix} \cdot \begin{pmatrix} x_r \\ x_i \end{pmatrix} = \begin{pmatrix} 19x_r - 43x_i \\ 43x_r + 19x_i \end{pmatrix} . \qquad (2.5)$$

One solution to realize this matrix multiplication is to use the MCM circuit of Figure 2.2(c) two times for real and imaginary input and to add/subtract the results as illustrated in Figure 2.3(a). Even if the MCM solutions are optimal in terms of number of adders, the CMM circuit in Figure 2.3(a) is not optimal as there exist a CMM circuit with two adders less as shown in Figure 2.3(b). Hence, redundancies between different inputs were used to reduce the complexity. Finding an adder circuit with minimum adders for a given constant matrix is known as the CMM problem.

2.4 Definitions

Before starting to formulate the optimization problems and discussing the previous work on how to solve these, some common concepts and notations

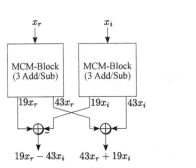

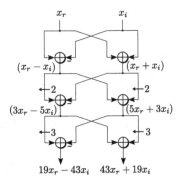

(a) CMM with two MCM operations   (b) Optimized CMM operation

Figure 2.3: Example CMM operation with constant $19 + j43$

that are used throughout this work are introduced. The notation is close to the framework introduced by Voronenko and Püschel [41] which unifies many concepts from previous work.

2.4.1 Adder Graph Representation of Constant Multiplication

The adder circuits for constant multiplication as introduced above can be represented as a directed acyclic graph (DAG) [42] which is called *adder graph* in the following. Each node except the input node corresponds to an adder and has an in-degree of two. Each edge weight corresponds to a bit shift. An integer number can be assigned to each node of the graph which corresponds to the multiple of the input. This value is commonly called a *fundamental* [42]. The operation of each node can be fused into a single generalized add/subtract operation which is called $\mathcal{A}$-operation [41]:

Definition 1 ($\mathcal{A}$-operation). *An $\mathcal{A}$-operation has two input fundamentals $u, v \in \mathbb{N}$ and produces an output fundamental*

$$\mathcal{A}_q(u, v) = |2^{l_u} u + (-1)^{s_v} 2^{l_v} v| 2^{-r} \qquad (2.6)$$

where $q = (l_u, l_v, r, s_v)$ is a configuration vector which determines the left shifts l_u, $l_v \in \mathbb{N}_0$ of the inputs, an output right-shift $r \in \mathbb{N}_0$ and a sign bit $s_v \in \{0, 1\}$ which denotes whether an addition or subtraction is performed.

Note that output right shifts ($r > 0$) can alternatively be represented by allowing negative values for the input shifts, e.g., $l'_{u/v} = l_{u/v} - r$ ($l'_{u/v} \in \mathbb{Z}_0$), although it will be implemented differently in hardware. The adder graph of the circuit in Figure 2.2(c) using this representation is illustrated in Figure 2.4. The node values correspond to the fundamentals $w = \mathcal{A}_q(u, v)$, where u and v are the nodes connected at the input, the sign (s_v) is indicated by '+' or '−' and $l'_{u/v}$ are used as edge weights. Node '1' corresponds to the input terminal. For example, node '5' is realized by left shifting the input x by 2 bits and adding the unshifted input, leading to $2^2 x + 2^0 x = 5x$. Note that no explicit right shift r is used in this representation as it can be easily derived from negative edge values.

It was shown by Dempster and Macleod that any adder graph can be transformed to an unique adder graph with identical topology (and hence identical adder count) where each fundamental is represented by an odd fundamental, which they called an *odd fundamental graph* [42]. In addition, the MCM search can be typically restricted to only positive fundamentals without limiting the search space as they can be negated by changing adders to subtractors (of the adder graph or succeeding adders) and vice versa [42, 43]. Hence, given a set of target constants T to be optimized, the first step in any MCM algorithm is to compute the unique odd representation of their magnitudes

$$T_{\mathrm{odd}} := \{\mathrm{odd}(|t|) \mid t \in T\} \setminus \{0\}, \tag{2.7}$$

denoted as the target set, where $\mathrm{odd}(t)$ is equal to t divided by 2 until it is odd. Throughout this work, adder graphs are always assumed to be odd fundamental graphs.

2.4.2 Properties of the $\mathcal{A}$-Operation

In the following, some useful properties of the $\mathcal{A}$-operation are derived which are used in this work and follow the derivations from [41].

Lemma 1. *If there exists a configuration q with $w = \mathcal{A}_q(u, v)$ then there also exists a valid configuration q' for which $w = \mathcal{A}_{q'}(v, u)$.*

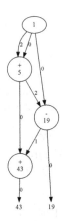

Figure 2.4: Adder graph realizing the multiplication with the coefficients $\{19, 43\}$

Proof. The relation $|a \pm b| = |b \pm a|$ can be used to rearrange (2.6) as follows

$$
\begin{aligned}
w &= \mathcal{A}_q(u, v) \\
&= |2^{l_u} u + (-1)^{s_v} 2^{l_v} v| 2^{-r} \\
&= |2^{l_v} v + (-1)^{s_v} 2^{l_u} u| 2^{-r} \\
&= \mathcal{A}_{q'}(v, u)
\end{aligned}
\tag{2.8}
$$

with $q' = (l_v, l_u, r, s_v)$. □

Lemma 2. *If there exists a configuration q with $w = \mathcal{A}_q(u, v)$ then there also exists a valid configuration q'' for which $u = \mathcal{A}_{q''}(w, v)$.*

Proof. Rearranging (2.6) to u results in

$$
u = \begin{cases}
(2^r w + (-1)^{\overline{s_v}} 2^{l_v} v) 2^{-l_u} & \text{when } 2^{l_u} u + (-1)^{s_v} 2^{l_v} v \geq 0 \\
-(2^r w + (-1)^{s_v} 2^{l_v} v) 2^{-l_u} & \text{otherwise}
\end{cases}
\tag{2.9}
$$

leading to

$$
|u| = A_{q''}(w, v)
\tag{2.10}
$$

with $q'' = (r, l_v, l_u, s_v')$ and

$$
s_v' = \begin{cases}
\overline{s_v} & \text{when } 2^{l_u} u + (-1)^{s_v} 2^{l_v} v \geq 0 \\
s_v & \text{otherwise}
\end{cases}
\tag{2.11}
$$

□

2.4.3 $\mathcal{A}$-Sets

Besides the $\mathcal{A}$-operation, it is useful to define a set which contains all possible numbers that can be obtained from u and v by using one $\mathcal{A}$-operation:

Definition 2 ($\mathcal{A}_*(u,v)$-set). *The $\mathcal{A}_*(u,v)$-set contains all possible fundamentals which can be obtained from u and v by using exactly one $\mathcal{A}$-operation:*

$$\mathcal{A}_*(u,v) := \{\mathcal{A}_q(u,v) \mid q \text{ is a valid configuration}\} \qquad (2.12)$$

A valid configuration is a combination of l_u, l_v, r and s_v such that the result is a positive odd integer $\mathcal{A}_q(u,v) \leq c_{\max}$.

The limit $c_{\max}$ has to be introduced to keep the set finite. It is usually chosen as power-of-two value which is set to the maximum bit width of the target values plus one [41, 44],

$$c_{\max} := 2^{B_T+1} \qquad (2.13)$$

with

$$B_T = \max_{t \in T_{\text{odd}}} \lceil \log_2(t) \rceil . \qquad (2.14)$$

For convenience, the $\mathcal{A}_*$-set is also defined for two input sets $U, V \subseteq \mathbb{N}$, which contains all fundamentals that can be computed from the sets U and V by a single $\mathcal{A}$-operation:

$$\mathcal{A}_*(U,V) := \bigcup_{u \in U, \ v \in V} \mathcal{A}_*(u,v) \qquad (2.15)$$

In addition, the $\mathcal{A}_*$-set is also defined for single input set $X \subseteq \mathbb{N}$ as follows:

$$\mathcal{A}_*(X) := \bigcup_{u,v \in X} \mathcal{A}_*(u,v) \qquad (2.16)$$

Note that $\mathcal{A}_*(\{u,v\})$ is different from $\mathcal{A}_*(u,v)$ as the first one contains all combinations of u and v: $\mathcal{A}_*(\{u,v\}) = \mathcal{A}_*(u,u) \cup \mathcal{A}_*(u,v) \cup \mathcal{A}_*(v,v)$.

2.4.4 The MCM Problem

Using the graph representation above, the MCM problem can be stated as follows:

Definition 3 (MCM Problem). *Given a set of positive odd target constants T, find a valid adder graph with minimum adders that realizes the multiplication with each constant of T.*

A *valid* adder graph is given if there exists a path from the input node '1' to each of the target nodes in T_{odd}. With the definition of the $\mathcal{A}$-operation in (2.6) and the odd representation (2.7) we can separate the core of the MCM problem formally as follows (according to [41]):

Definition 4 (Core MCM Problem). *Given a set of positive odd target constants $T_{odd} = \{t_1, \ldots, t_M\}$, find the smallest set $R = \{r_0, r_1, \ldots, r_K\}$ with $T_{odd} \subseteq R$ and $r_0 = 1$ for which the elements in R can be sorted in such a way that for all triplets $r_i, r_j, r_k \in R$ with $r_k \neq r_i$, $r_k \neq r_j$ and $0 \leq i, j < k$, there is an $\mathcal{A}$-configuration p_k that fulfills:*

$$r_k = \mathcal{A}_{p_k}(r_i, r_j) \ . \tag{2.17}$$

In other words, the core of the MCM problem is to find the smallest intermediate set I such that a *valid* adder graph can be constructed from the nodes in $R = I \cup T_{odd}$. The elements from set I are denoted as non-output fundamentals (NOFs). When this set is found, it is a non-complex task to find the corresponding $\mathcal{A}$-configurations to build the graph. This can be done with the optimal part of the heuristics explained later in Section 2.6.2. Note that besides the minimum node count (i. e., number of adders), there also exist other cost metrics which employ the actual hardware costs in terms of full-adder cells or gate count [45–48].

2.4.5 MCM with Adder Depth Constraints

An important property of an adder graph is the so-called *adder depth (AD)* (sometimes referred to as *logic depth* or *adder-steps*). The *adder depth of node c*, denoted AD(c), is defined as the maximum number of cascaded adders on each path from input node '1' to node c. The minimum possible adder depth of node c, denoted $\mathrm{AD_{min}}(c)$, can be obtained by using a binary tree of adders from the CSD representation of c [49]. It can be directly computed from the constant by

$$\mathrm{AD_{min}}(c) = \lceil \log_2(\mathrm{nz}(c)) \rceil \tag{2.18}$$

where $\mathrm{nz}(c)$ denotes the number of non-zeros of the CSD representation of c. Obviously, an adder graph with minimal AD may need more adders than the solution with minimal number of adders.

The adder depth plays an important role for the delay and the energy consumption of an MCM circuit [34, 50–55]. Thus, it is useful for many applications to limit the adder depth besides the minimization of adders. To reduce the overall delay it is sufficient to limit the worst-case adder depth of all nodes. The lowest possible adder depth is limited by the node(s) with maximum adder depth

$$D_{\max} = \max_{t \in T} AD_{\min}(t) \ . \tag{2.19}$$

Limiting the adder depth to $D_{\max}$ leads to the lowest possible delay and to the corresponding optimization problem:

Definition 5 (MCM with Bounded Adder Depth (MCM$_{BAD}$) Problem). *Given a set of positive target constants $T = \{t_1, \ldots, t_M\}$, find an adder graph with minimum cost such that $AD(t) \leq D_{\max}$ for all $t \in T$.*

The adder depth has an impact on the power consumption [34, 50–55]. It is based on the insight that a transition which is generated at the input or an adder output produces more transitions (glitches) in the following stages. If the adder depth is larger than needed, this produces more transitions than needed and, thus, a higher dynamic power consumption. Therefore, it is desirable for low power applications that *each* output node in the graph is obtained with minimal AD, which is defined as follows:

Definition 6 (MCM with Minimal Adder Depth (MCM$_{MAD}$) Problem). *Given a set of positive target constants $T = \{t_1, \ldots, t_M\}$, find an adder graph with minimum cost such that $AD(t) = AD_{\min}(t)$ for all $t \in T$.*

2.5 Field Programmable Gate Arrays

In this section, a brief overview of the architecture of FPGAs is given, as needed to introduce the mapping of the aforementioned adder graphs. More details about FPGA fundamentals can be found in textbooks (e. g., [56,57]), an introduction to the arithmetic features of more recent FPGAs is given in the thesis of Bogdan Pasca [58].

The first commercial FPGA in its current form was introduced by the XC2000 series of Xilinx Inc. in 1985 [56]. Its main difference compared to its programmable logic predecessors is the array-like organization of programmable logic elements (LEs) which are embedded in a programmable

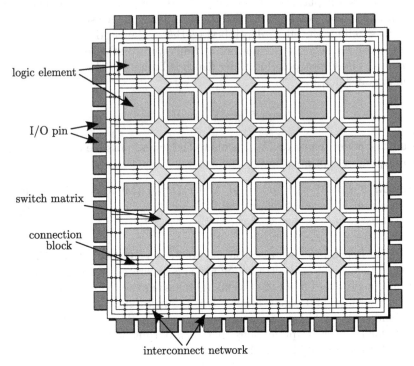

logic element

I/O pin

switch matrix

connection
block

interconnect network

Figure 2.5: Simplified architecture of a generic FPGA layout

interconnect network consisting of switch matrices and connection blocks as
indicated in Figure 2.5. The smallest common piece of logic of todays FPGAs
typically consists of a look-up table (LUT), carry-chain logic and flip flops
which is denoted as basic logic element (BLE) in the following. Each LE
is typically built from a cluster of BLEs, i.e., several BLEs are combined
with some local interconnect, to limit the input count. The interconnect or
routing network allows the combination of several LEs to build much more
complex logical circuits whose size is only limited by the resources of the
FPGA. As both LEs and the interconnect network, are configurable at runtime, the logical circuit can be defined after manufacturing the chip. Due
to this flexibility and low non-recurring engineering (NRE) cost, FPGAs became very popular as an alternative to ASICs. The huge overhead due to
the programmability is partly compensated by the smallest and fastest technologies which are only affordable in very high volume productions. Hence,

todays FPGAs are reaching a complexity in the order of millions of BLEs and a speed close to one GHz.

While all FPGAs have an array-like structure, they have many differences in the detailed routing and LE structure. In the next two subsections, the LE structure is further detailed for recent commercial high-end FPGA devices from the two largest vendors: Xilinx and Altera. Of course, this limited number of FPGA architectures can not be comprehensive but gives a reasonable overview about the state-of-the-art.

2.5.1 Basic Logic Elements of Xilinx FPGAs

On Xilinx FPGAs, the LE corresponds to a complex logic block (CLB) which contains several sub-units which are called slices. The Virtex 4 architecture contains four slices per CLB. Each Virtex 4 slice contains two BLEs consisting of a 4-input LUT, a carry-chain logic and a single flip flop (FF) [59]. The simplified structure of a Virtex 4 BLE is shown in Figure 2.6(a). It can realize any Boolean function of four input variables (a-d), followed by an optional register (selected by the multiplexer (MUX) close to the output y). Alternatively, the dedicated carry logic consisting of an exclusive-or (XOR) gate and a MUX can be used to build a fast full adder with dedicated carry-in (c_i) and carry-out (c_o) connections to other BLEs to build a ripple carry adder (RCA). For this, the LUT has to be configured as XOR gate between b and c and one of these ports has to be connected to the zero-input of the MUX. The AND gate can be used for generating partial products as needed in generic multiplications.

Starting from the Virtex 5 architecture, the LUT was extended to a 6-input LUT per BLE which can be configured into two independent 5-input LUTs that share the same inputs as shown in Figure 2.6(b) [60]. Each slice now contains four identical BLEs. In the Virtex 6, the Spartan 6 and the latest 7 series FPGA architectures, an additional FF was added (shown gray in Figure 2.6(b)) [61–63]. They also contain the same carry logic as the Virtex 4 but the LUT bypass and the AND gate are now realized using the second 5-input LUT.

2.5.2 Basic Logic Elements of Altera FPGAs

Recent Altera Stratix FPGAs are organized in logic array blocks (LABs) which contain sub-units called ALM. The Stratix III to Stratix V FPGAs

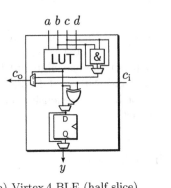

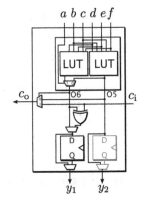

(a) Virtex 4 BLE (half slice)   (b) Virtex 5/6, Spartan 6 and 7 Series
                                         BLE (quater slice)

Figure 2.6: Simplified BLE of Xilinx FPGA families

contain ten ALMs per LAB [64–66]. The functionality of a BLE as defined
above is roughly covered by one half of an ALM. The structure of this Altera
BLE mainly consists of one 4-input LUT, two 3-input LUTs, a dedicated FA
and two FFs as shown in Figure 2.7. The second FF (shown gray in Figure 2.7) was introduced with the Stratix V architecture and is missing in
older Stratix FPGAs. Each ALM has eight input lines which have to be
shared between two BLEs. Multiplexers between both BLEs in the same
ALM can be used to adapt the logic granularity to different effective LUT
sizes (which is called adaptive LUT (ALUT)). The possible configurations
of a complete ALM are two independent 4-input LUTs, two independent
3-input and 5-input LUTs, one independent 6-input LUT, various combinations of LUTs up to six inputs where some of the inputs are shared and some
specific 7-input LUTs [64–66].

2.5.3 Mapping of Adder Graphs to FPGAs

As mentioned above, BLEs of modern FPGAs directly contain dedicated
logic and routing resources to build fast ripple carry adders (RCAs). An
example configuration of a small 4-bit RCA using the CLBs of a Xilinx
Virtex 4 FPGA is given in Figure 2.8(a). Each BLE computes the Boolean

$a\ b\ c\ d\ e$

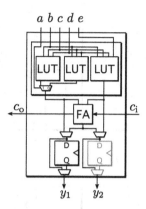

Figure 2.7: Simplified BLE (half ALM) of Altera Stratix III to Stratix IV FPGAs

relations

$$s_k = a_k \oplus b_k \oplus c_k \tag{2.20}$$

$$c_{k+1} = (a_k \oplus b_k)c_k + \overline{a_k \oplus b_k}a_k \tag{2.21}$$

$$= a_k c_k + b_k c_k + a_k b_k \tag{2.22}$$

$$\tag{2.23}$$

which correspond to the function of a full adder (FA).

Each RCA is pretty fast due to the fast carry chain multiplexers. An RCA with B_o output bits is compact and its critical path, which is typically from the LSB input to the most significant bit (MSB) output, consists of the delay of one LUT (denoted τ_{LUT}), $B_o - 1$ times the carry-chain of a single BLE (denoted τ_{CC}), and one XOR gate (denoted τ_{XOR}) [67]:

$$\tau_{\text{RCA}}(B_o) = \tau_{\text{LUT}} + (B_o - 1)\tau_{\text{CC}} + \tau_{\text{XOR}} \tag{2.24}$$

To give some example values, consider the Virtex 6 FPGA architecture. They are available with different performance grades which are called *speed grades*. The carry-in to carry-out delay of a complete slice with four BLEs for the fastest speed grade of 3 is specified to be at most 0.06 ns [68]. Hence, each BLE contributes with $\tau_{\text{CC}} = 0.015$ ns to the carry delay. The relevant combinational delay of a LUT ranges between $\tau_{\text{LUT}} = 0.14\ldots0.32$ ns depending which input influences which output [68]. The delay of the XOR (from carry-in to one of the BLE outputs within a slice) is $\tau_{\text{XOR}} = 0.21\ldots0.25$ ns. For this example, the worst-case delays of 8, 16 and a 32 bit adders are about 1 ns, 1.5 ns and 2.5 ns, respectively.

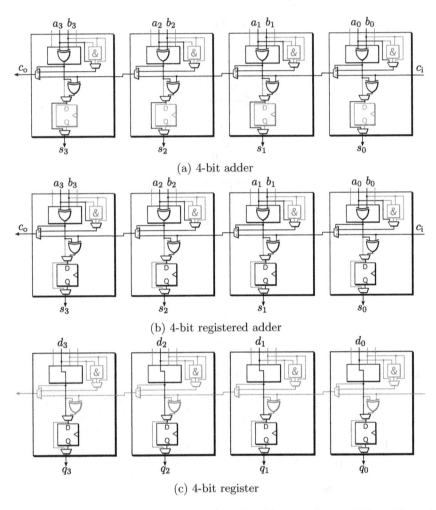

(a) 4-bit adder

(b) 4-bit registered adder

(c) 4-bit register

Figure 2.8: BLE configurations for pipelined adder graphs on Xilinx Virtex 4
FPGAs

Note that it is well known that in general VLSI designs the RCA is the
most compact adder structure but also the slowest one [69, 70]. Much faster
adder structures are known but most of them are counter productive on
FPGAs due to the dedicated fast carry logic. The only exception are large
word lengths of at least 32 bit and more [67, 71, 72].

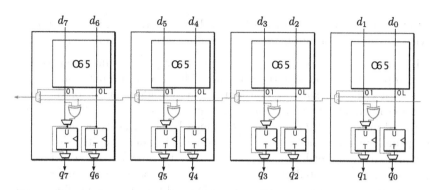

Figure 2.9: BLE configuration for an 8 bit register on Xilinx Virtex 6 FPGAs

Clearly, each node of an adder graph can be mapped to such an RCA structure. While in VLSI realizations the delay of local wires can often be neglected, a connection between two LEs on an FPGA typically includes several routing elements which significantly affects the overall delay of the circuit. A typical local routing delay on a Virtex 6 FPGA is in the order of $0.3 \ldots 0.5$ ns (obtained from timing analyses of representative designs). Thus, a one-to-one mapping of an adder graph to BLEs accumulates the adder and routing delays with the consequence that the speed potential of the FPGA is by far not exhausted. The common countermeasure to break down the critical part is pipelining [73]. To give an example of the potential speed, carefully pipelined designs can run at 600 MHz and above on a Virtex 6 FPGA. Here, even a single local routing connection in the critical path consumes about one third of the clock period of 1.66 ns.

Pipelining of non-recursive circuits like the adder circuits under consideration is done by breaking the combinational paths and placing additional FFs in between such that each resulting path from an input to an output has the same number of pipeline FFs [73]. The number of FFs in each path from input(s) to output(s) is called the *pipeline depth* of the circuit. Pipelining only increases the latency by the pipeline depth and keeps the functionality of the circuit. Pipelining an RCA on an FPGA is cost neutral as the otherwise unused FFs are simply used as shown in Figure 2.8(b).

However, to guarantee that each path has the same number of pipeline FFs, additional registers have to be placed to balance the pipeline. The BLE configuration of a pure register with 4 bit is shown in Figure 2.8(c). Clearly, on this FPGA architecture, a register consumes exactly the same resources in terms of BLE resources as an adder of the same word size (combinational or registered). On modern FPGAs each BLE can realize two FF leading to twice the number of bits for registers as illustrated in Figure 2.9.

However, as many of these pipeline balancing registers may be necessary, it is important to consider them in the adder graph optimization. To give a first motivating example, consider again the adder graph of the MCM solution for multiplying with the constants $\{19, 43\}$ of Figure 2.4 which is repeated in Figure 2.10(a). It was obtained by the H_{cub} algorithm [41] and is optimal in terms of the number of required adders (as its adder count is equal to the lower-bound given in [74]). To pipeline this adder graph, each adder is augmented by a pipeline register and three additional registers are required to balance the pipeline. This is illustrated as pipelined adder graph (PAG) in Figure 2.10(b). Here, each node with a rectangular shape includes a pipeline register, i. e., two-input nodes marked with '+' or '–' are pipelined adders and subtractors, respectively, while single input nodes are pure registers. Each node in the PAG corresponds to similar BLE costs. As illustrated in Figure 2.10(c), there obviously exists a solution with less nodes. This PAG will typically use less BLEs although more adders are used.

Note that optimizing PAGs is complementary to methods that increase the speed of a single adder, e. g., deeply pipelined adders [67]. In contrast to this, some registers can be saved and the latency can be reduced at the cost of a reduced throughput by skipping some of the stages. Both cases can be considered by adjusting the cost of the registers and adders in each stage, the core optimization problem remains the same (the details are elaborated in Section 5.4).

2.5.4 Cost Models for Pipelined Adders and Registers

Assuming that each BLE is able to realize a single FA, a single FF or a combination of both, a pipelined RCA with B_o output bits requires exactly B_o BLEs. The output word size of a node computing a factor w of the input requires $\lceil \log_2(w) \rceil$ bits in addition to the input word size of the MCM block. Thus, the cost in terms of the number of BLEs for a *pipelined* $\mathcal{A}$-operation can be obtained by

$$\text{cost}_A(u, v, w) = B_o = B_i + \lceil \log_2(w) \rceil + r \qquad (2.25)$$

where B_i is the input word size of the MCM block and $r \in q$ is the right shift required to obtain $w = \mathcal{A}_q(u, v)$ according to (2.6). Note that there may be a significant difference in the cost compared to the *non-pipelined* $\mathcal{A}$-operation as FFs can be replaced by wires.

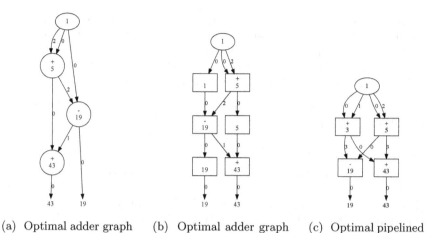

(a) Optimal adder graph   (b) Optimal adder graph   (c) Optimal pipelined
                              after pipelining   adder graph

Figure 2.10: Adder graphs realizing the multiplication with the coefficients $\{19, 43\}$

The right shift r has to be considered in the cost as additional half adders for the carry propagation of LSBs are required if right shifts are used. Take for example the computation of $5x = (3x + 7x)2^{-1} = 10x2^{-1}$ using $r = 1$. It will require $\lceil \log_2(10) \rceil = 4$ additional bits (on top of B_i) for its computation but only $\lceil \log_2(5) \rceil = 3$ additional bits are used to store the right shifted result. As right shifts occur relatively seldom and it takes effort to compute the right shift value for the cost evaluation, a good approximation is obtained by ignoring possible right shifts:

$$\text{cost}_A(w) = B_i + \lceil \log_2(w) \rceil \tag{2.26}$$

The register costs are modeled in a similar way as again $B_i + \lceil \log_2(w) \rceil$ FF are used for storage:

$$\text{cost}_R(w) = (B_i + \lceil \log_2(w) \rceil) / K_{FF} \tag{2.27}$$

Here, K_{FF} represents the number of FFs per BLE. Throughout the thesis, this cost model at BLE level is called the *low-level* cost model. Simply counting the adders and pure registers (i.e., $\text{cost}_A(u, v, w) = 1$ and $\text{cost}_R(w) = 1$) is called the *high-level* cost model. Note that the low-level model described above is different from the number of FAs compared to previous work on non-pipelined adder graphs [45–48] as additional BLEs are required for FFs in the pipelined case. The cost of the total adder graph, denoted as cost$_{PAG}$, corresponds to the sum of the cost values of each node. Note that some FPGA vendors provide efficient register chains that use the flip flops that usually store the LUT content in a very efficient way (SRL16/SRLC32E primitives on Xilinx FPGAs [75]). This will be considered in detail in Section 5.4.

For ASICs, a good low-level approximation for RCA-based PAG implementations is given by

$$\text{cost}_R(w) = c_{FF}(B_i + \lceil \log_2(w) \rceil) \tag{2.28}$$

$$\text{cost}_A(w) = (c_{FA} + c_{FF})(B_i + \lceil \log_2(w) \rceil) \tag{2.29}$$

where c_{FF} and c_{FA} correspond to the area cost of a single FF and FA, respectively. The number of FAs might be slightly larger compared to the more sophisticated models [45–48] but gives a good approximation.

2.6 Related Work

The related work of SCM and MCM is given in the following. Related work to specific topics is discussed in the corresponding section when necessary.

2.6.1 Single Constant Multiplication

The idea to perform the multiplication as a sum of power-of-two values is pretty old. The first known documentation of such a scheme is the mathematical Papyri Rhind of Ahmes which dates back to the seventeenth century B.C. The method is known as the Egyptian multiplication (sometimes also referred to as Russian multiplication) [76]. One operand is decomposed into a sum of power-of-two values by repeated subtraction of the largest power-of-two. The other operand is duplicated multiple times and the corresponding duplicates are selected and added. Take, for example, the multiplication of 77 times 89. The largest power-of-two term less than 77 is 64, the next one less than $77 - 64 = 13$ is 8, etc., leading to $77 = 64 + 8 + 4 + 1 = 2^6 + 2^3 + 2^2 + 2^0$. Then, the other operand is duplicated by multiple additions, e. g., $2 \cdot 89 = 89 + 89 = 178$, $4 \cdot 89 = 178 + 178 = 356$, $8 \cdot 89 = 356 + 356 = 712$, etc. The multiplication result is obtained by adding the multiples corresponding to the power-of-two decomposition of the first operand, i. e., $77 \cdot 89 = (64 + 8 + 4 + 1) \cdot 89 = 5696 + 712 + 356 + 89 = 6853$.

One can say that the ancient Egyptians implicitly used the binary number system which is nowadays the base of most digital arithmetic.

The use of signed digits to simplify multiplication as well as division can be dated back to 1726 [77]. Its efficient use in digital systems became popular in the 1950'th [26–28]. Later, the complexity was further reduced by the use of common subexpression elimination (CSE) of repeated patterns in the number representation [12,33]. An additional step towards fewer complexity was the graph based representation of constant multiplications as used in digital filters [78]. Adder graph based representations yield to optimal SCM circuits (in terms of the number of additions) which were exploited by the minimized adder graph (MAG) algorithm of Dempster and Macleod for up to four adders covering all 12 bit coefficients [42]. Later, these results were enhanced by introducing simplifications by Gustafsson et al. to five adders covering all coefficients up to 19 bit [79, 80]. Further extensions for up to six adders resulted in optimal SCM circuits for coefficients up to 32 bit [81, 82]. In the latter work, not all graphs are generated exhaustively but the minimal representation of a given constant is obtained by the graph topology tests introduced by Voronenko and Püschel [41]. Although it was shown by Cappello and Steiglitz that the SCM problem is NP-complete [83], the SCM problem can be regarded as solved for most of the practical applications (which typically include coefficients with less than 32 bit).

2.6.2 Multiple Constant Multiplication

The MCM problem is a generalization of the SCM problem and has been an active research topic for the last two decades. Many different optimization methods have been proposed [7, 9, 12, 33, 41, 43–45, 48, 49, 78, 84–101]. There are two fundamentally different concepts for solving the MCM problem, in the first concept, intermediate values are obtained from common subexpressions [12, 33, 89, 92–94], in the second concept, graph based methods are used [7, 9, 41, 43, 44, 49, 78, 84–91, 95–101]. CSE methods search for identical patterns in the binary or signed digit number representation of the constant(s). As demonstrated in Section 2.1.3, this method may fail to find a minimal solution due to hidden non-zeros [34, 102]. Some of these hidden non-zeros can be extracted using the method proposed by Thong and Nicolici but some of them may remain hidden [102].

The hidden non-zero problem is inherently avoided in graph based methods as they are not restricted to the number representation. A detailed comparison between optimal graph based methods and CSE with different

number representations (without considering hidden non-zeros) can be found
in the work of Aksoy et al. [99]. On average, 10.8 additional adders were required for an exact CSE-based solution compared to the exact graph based
solution in an experiment with 30 MCM instances with 10 to 100 random
coefficients with 14 bit each. However, as the search space is much larger for
graph based methods, it can still be advantageous to use CSE methods for
large problems [103].

As the MCM problem is a generalization of the SCM problem it is also
NP-complete. Hence, even if there exist optimal methods, there is a need of
good heuristics for larger problem sizes. The n-dimensional reduced adder
graph (RAG-n) algorithm, introduced by Dempster and Macleod [44], was
one of the leading MCM heuristics for many years. This situation changed
significantly with the introduction of the H_{cub} algorithm by Voronenko and
Püschel [41] and the introduction of the difference adder graph (DiffAG)
algorithm by Gustafsson [43], both in 2007. Both state-of-the-art MCM
heuristics are briefly introduced in the following.

The H_{cub} algorithm starts, like most other MCM algorithms, with an algorithm that is called the *optimal part* of the MCM optimization. A pseudo-
code of this part is given in Algorithm 2.1. The input to the algorithm is the
target set T containing all coefficients. First, the unique odd representation
is computed (line 2). Then, the so-called *realized set* R is initialized to a
set containing only the element '1'. Next, the so-called *successor set* S is
computed, which contains all odd fundamentals which can be computed by
one addition from the realized set (line 5). It is now checked if new target
elements can be realized and the corresponding target elements are stored
in set T' (line 6). These elements (if any) are inserted into R and removed
from T (lines 7 and 8). This procedure is repeated until no further target
element can be realized (then, T' is empty). If the target set T is also empty,
it is known that the solution is optimal (each target can be computed by one
adder or subtractor). If elements in T remain, at least one additional NOF
from the successor set has to be inserted into R to compute the remaining
elements of T. For that, adder graph topologies with up to three adders are
evaluated in the H_{cub} algorithm to obtain or estimate (in case that more
than three additional adders are necessary) the so-called $\mathcal{A}$-distance of all
possible successors in S. The $\mathcal{A}$-distance (which is equivalent to the adder
distance [44]), denoted as $\text{dist}(R, c)$, is defined as the minimum number of
$\mathcal{A}$-operations necessary to compute c from R. The main idea is to select the

Algorithm 2.1: The optimal part of the MCM optimization

```
1  R = MCM_opt (T)
2     T ← {odd(|t|) | t ∈ T} \ {0}
3     R := {1}
4     do
5        S ← A_*(R)
6        T' ← T ∩ S
7        R ← R ∪ T'
8        T ← T \ T'
9     while T' ≠ ∅
```

successor s leading to the best benefit

$$B(R, s, t) = \text{dist}(R, t) - \text{dist}(R \cup \{s\}, t) \tag{2.30}$$

for all remaining target coefficients $t \in T$ (in fact, a slightly modified *weighted* benefit function is used [41]). The best successor s is included in R and the optimal part of the algorithm is evaluated again which moves elements from T to R. This procedure is continued until T is empty. The implementation of the H_{cub} algorithm is available online as source code [104]. This implementation includes a switch to solve the MCM problem with bounded adder depth, denoted as $H_{\text{cub,BAD}}$ in the following. The idea of H_{cub} was later extended in the H3 and H4 heuristics by Thong and Nicolici [100]. The key improvements were obtained by computing the adder distance exactly for H4 and using better estimators for larger distances in both heuristics. With that they obtained better results and less runtime (due to a restriction of c_{max}) compared to H_{cub} for problems with larger word length and low coefficient count.

The DiffAG algorithm [43] follows a different strategy. It basically starts with the optimal part as given in Algorithm 2.1. Then, an undirected complete graph is formed where each node corresponds to a set of coefficients, initially one node for the realized set $N_0 = R$ and for each remaining target element, one set containing the target element $N_i = \{t_i\}$ for all $t_i \in T_{\text{odd}} \setminus R$. Now, the so-called *differences* are computed between each pair of nodes in the graph and are inserted into the difference sets D_{ij}. A difference $d_{ij} \in D_{ij}$ is a fundamental that allows – when inserted into R – to compute the elements in N_i from the elements in N_j and vice versa. Each edge (i, j) corresponds to one set of differences D_{ij}. Formally, for each $n_i \in N_i$ and $n_j \in N_j$, there exists an A-operation such that $n_i = A_q(n_j, d_{ij})$ and, due to Lemma 2 (see page 16), $n_j = \mathcal{A}_{q'}(n_i, d_{ij})$. The difference set can be directly computed from the node values by $D_{ij} = \mathcal{A}_*(N_i, N_j)$. If there exists a difference which is included in the realized set ($d_{ij} \in R$), the corresponding nodes are merged into a single node $N_k = N_i \cup N_j$, N_i and N_j are removed and the corresponding edges and difference sets are merged. If no difference remains that is included in the realized set, the idea is to realize the least-cost and most frequent difference. This process is continued until all nodes in the graph are merged into a single node which set corresponds to the final solution. The results in [43] show that the DiffAG algorithm is advantageous for large target sets with low coefficient word size compared to the H_{cub} algorithm.

Besides the heuristic approaches discussed above, a lot of research effort has been taken towards optimal MCM methods in the last years [9, 96–100]. An ILP-based method for finding the common subexpressions in an MCM problem was proposed by Aksoy et al. [96]. However, as mentioned above, this will not necessarily lead to the least adder solution due to the number representation dependency. Another approach for an optimal MCM method using ILP was proposed by Gustafsson [97]. Here, the MCM problem is formulated as the problem of finding a Steiner hypertree in a directed hypergraph. A hypergraph is an extended graph, where each edge, called hyperedge, can connect more than two nodes. A Steiner hypertree is a straight forward extension of the Steiner tree, which is defined as a minimum weight tree spanning a given set of vertices using some additional vertices (if required), to hypergraphs [105]. The details of this approach will be discussed in Section 5.1. It finds optimal solutions and can be adapted to several additional MCM constraints like minimal adder depth or fanout restrictions. However, due to its computational complexity it is only suitable for small MCM instances or to find better lower bounds (e. g., compared to [74, 106]) by relaxing the model to a continuous LP problem. Optimal breadth-first search and depth-first search algorithms based on branch-and-bound for the graph-based MCM problem were proposed by Aksoy et al. [9, 98, 99], leading to the first optimal MCM method for real-world FIR filter applications. A similar approach that can handle problems with larger word lengths but less coefficients was used by the bounded depth-first search (algorithm) (BDFS) algorithm introduced by Thong and Nicolici [100].

2.6.3 Modified Optimization Goals in SCM and MCM Optimization

Besides the reduction of adders in the SCM/MCM problem, attention was paid to different optimization goals related to the circuit implementation in more recent work. As already mentioned above, the power consumption is related to the adder depth of the adder graph, so many optimization algorithms solving the MCM_{MAD} problem were proposed [34, 50–55, 107–109]. Besides the adder depth, the glitch path count (GPC) was proposed as a more accurate measure to estimate a power equivalent at high level [50]. Here, different paths from the input to each adder are counted which is approximately proportional to the dynamic power. The GPC power estimation was further refined to the glitch path score (GPS) [52].

Another important optimization goal that considers the bit-level cost was first proposed by Johansson et al. [45, 46] and later used by Aksoy et al. [47] and Brisebarre et al. [48]. They observed that in computations like $w = 2^{l_u}u + v$, the l_u lowest bits of w are identical to the l_u lowest bits of v and, hence, a lower number of full adders is sufficient. Their models consider the bit shifts of all the cases that can occur in an adder graph leading to an accurate number of full adders, half adders and even inverters [47]. A bit-level model can also be used to reduce the critical path delay as demonstrated recently by Lou et al. [110, 111].

Besides carry propagate adders (CPAs) like the ripple-carry adders, carry-save adders (CSAs) can be used to reduce the delay as it is usually done in parallel multipliers. But the number of CSAs is different to the number of CPAs for the same adder graph as there is no 1:1 mapping possible (some adders can be replaced by wires while others require two CSAs). Hence, another modified optimization goal is to reduce the number of CSAs in MCM [112–115] or digital filters [116, 117].

2.6.4 Pipelined Multiple Constant Multiplication

Another method to increase the speed of a combinational circuit is pipelining [73]. It was mentioned in an early work of Hartley that *"it is important to consider not only the number of adders used, but also the number of pipeline latches, which is more difficult to determine"* [33]. As discussed earlier, this is in particular important on FPGAs as the routing delay is much larger compared to ASICs. Hence, Macpherson and Stewart proposed the reduced slice graph (RSG) algorithm [118] which is an FPGA-specific MCM optimization algorithm based on ideas of the RAG-n [44] and MAG [42] algorithms. Their aim is to reduce the adder depth as this reduces the number of additional registers to balance the pipeline. For that, they obtained optimal SCM solutions from the MAG algorithm but used the adder depth as primary selection metric and the adder count only as a secondary metric. Then, in contrast to RAG-n, non-trivial coefficients with highest cost are realized using the known SCM solution. Next, the optimal part of RAG-n (see Algorithm 2.1) is applied to find out if other coefficients can be realized using the new NOF. This procedure is repeated until all coefficients are realized. The solution typically has a larger adder count but a lower adder depth, leading to less registers after pipelining. They compared their results with parallel distributed arithmetic (DA) implementations which are often called to be perfectly suited for FPGAs as they are realized using LUTs and adder arithmetic only. They showed that considerable logic reductions are possible by using their method compared to DA.

The impact of adder graph pipelining was further quantified by Meyer-Baese et al. [119] who showed that average speedups of 111% can be achieved by a very moderate area increase of 6%. They used adder graphs obtained by the RAG-n algorithm [44] with a hand-optimized pipeline. They also compared their designs to parallel DA implementations and quantified the resource reductions to 71% on average.

Another optimization method that targets pipelined MCM circuits for FPGAs is the 'add and shift' method proposed by Mirzaei et al. [5, 6, 120]. They used a CSE algorithm to extract non-output fundamentals while keeping the AD minimal. Again, this reduces additional pipeline registers as a side effect. They reported average slices reductions of 50% compared to parallel DA.

An algorithm that already considers the cost of the pipeline registers during the optimization was first proposed by Aksoy et al. [7]. They used an iterative method, called HCUB-DC+ILP, where each iteration works as follows: First, an MCM solution is obtained by using the $H_{cub,BAD}$ algorithm with an iteration-specific random seed. The resulting coefficients (target and intermediate values) are inserted into the *realized set* R. In addition, all cost-1 coefficients of the form $2^k \pm 1$ up to the maximum word size of the target coefficients are also inserted into R. Then, all possible $\mathcal{A}$-operations to realize the target coefficients in R are computed and the best combination (in terms of gate-level area [121]) is searched using a binary integer linear programming (BILP) method which also considers the cost for pipeline registers. Finally, the pipeline method of [118] is applied and the costs are obtained and stored if they are better than any solution before. The next iteration starts with a different random seed to obtain a (possibly) different MCM solution and the results are added to R. The optimization stops when a fixed size of R is reached or if $H_{cub,BAD}$ does not find new solutions for a fixed number of iterations. Their target was a $0.18\mu m$ standard cell library for an ASIC design. They achieved a 5.6% area reduction on average compared to $H_{cub,BAD}$ solutions, which were automatically pipelined by the retiming of the logic synthesis tool (Cadence Encounter RTL Compiler). Compared to the non-pipelined solution (by removing the registers), the speedup is 34.0% which comes along with an area increase of 55.1% for the pipeline registers.

As a side effect, pipelining also reduces the logic depth and the power consumption of a circuit [122]. The pipeline registers "isolate" power producing glitches from one logic stage to the next yielding low power designs similar to designs obtained by the operand isolation technique introduced by Correale [123]. In MCM circuits, power consumption reductions of 43.0% due to pipelining were reported even though the area was increased [7]. The effect of the power reduction due to pipelining is expected to be even larger on FPGAs due to the programmable routing [124]. Wilton et al. [124] reported reductions between 40% and 90% in generic logic circuits. Hence, pipelining is a method to improve both figure of merit axes "performance" and "energy".

Part I

The Pipelined Multiple Constant Multiplication Problem

3 Optimal Pipelining of Precomputed Adder Graphs

Much effort was spent in previous work to solve the MCM problem as shown in the last chapter. The mapping of the corresponding adder graphs to an FPGA is straight forward as the FPGA already provides fast carry chains to build ripple carry adders. However, the direct mapping of the corresponding adder graph to an FPGA may lead to delays which are far beyond the speed capabilities of FPGAs. This speed can be greatly advanced by using pipelining. However, pipelining leads to additional resources which have to be considered. In this chapter, the focus is on pipelining an already optimized adder graph. It was originally published in [125] and serves as base for comparison with the direct PMCM optimization presented in Chapter 4 which improves the results obtained by the methods of this chapter.

3.1 Pipelining of Adder Graphs using Cut-Set Retiming

Valid pipelines can be formally obtained by using the cut-set retiming (CSR) method [73]. A cut-set is a set of edges that splits a graph G into two subgraphs G_1 and G_2 when all edges of the cut-set are removed from G. If all edges of the cut-set have the same direction, registers can be inserted into these edges resulting in a valid pipeline. Of course, the quality of the result strongly depends on how the cut-sets are chosen. An example of choosing cut-sets for each layer of adders in a straight forward way is illustrated by the dashed lines in Figure 3.1(a). Placing a register at each cut leads to a PAG as shown in Figure 3.1(b). As introduced in Section 2.5.3, each node with a rectangular shape includes a pipeline register. As the BLE cost for a single register is similar to that of a pipelined adder, the total cost of a PAG can be estimated by the number of registered operations (nodes in the PAG).

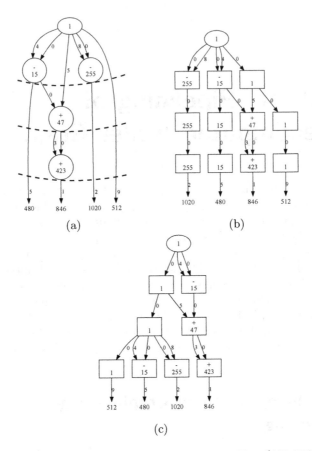

Figure 3.1: Adder graphs for the coefficient target set $T = \{480, 512, 846, 1020\}$, (a) adder graph without pipelining, (b) pipelined adder graph after cut-set retiming (c) optimized adder graph

The example PAG consists of four pipelined adders and seven pure registers resulting in eleven registered operations. Thus, the cost of the design is now dominated by the register cost. The assignment of operations to pipeline stages is very similar to a scheduling problem where each time step in the scheduling corresponds to a pipeline stage. Hence, an operation is said to be scheduled in stage s in the following when it is computed in pipeline stage s. The choice of the cut-sets directly corresponds to the scheduling of the adders. The scheduling of the operations in Figure 3.1(b) corresponds to an as soon as possible (ASAP) schedule [73].

By inspecting this example, the following observations can be made:

1. Node '255' can be scheduled at any pipeline stage as its input node '1' is available in any stage. Thus, moving the computation of node '255' to the last stage saves two registers.

2. Node '15' has to be computed in the first stage (to compute node '47'), it is not needed in the second stage but needed in the last stage. As its input node '1' is available in the second stage it *can* be computed again in the last stage and removed in the second stage. Thus, if the cost for two registers is larger than a single adder, it is better in terms of area to duplicate the adder in the last stage.

The result after applying these transformations is shown in Figure 3.1(c). Now, eight registered operations are needed instead of eleven operations of the graph in Figure 3.1(b). Finding the best scheduling as well as the adders which might be duplicated is in general as trivial as in the simple example above. Eliminating a register in one stage may prevent the reduction of several registers in later stages. Hence, a binary integer linear programming (BILP) formulation (also known as 0-1 ILP) for scheduling and adder duplication was developed which is described in the next section.

3.2 Minimized Pipeline Schedule using Integer Linear Programming

Several heuristic and optimal algorithms exist for the generic scheduling problem. For static scheduling, ILP (or BILP) is often used to optimally solve the problem [73]. In this section, we propose a BILP formulation for the optimal pipeline schedule of a given adder graph including the adder duplication mentioned above. It was originally published in [125] and is further extended and simplified in this work.

3.2.1 Basic ILP Formulation

Binary variables are used to describe if a node of the adder graph is scheduled in a given pipeline stage:

$$x_w^s = \begin{cases} 1 & \text{when node } w \text{ is scheduled in pipeline stage } s \\ 0 & \text{otherwise} \end{cases} \tag{3.1}$$

Each output of the adder graph is represented by the set of odd target fundamentals $T_{\text{odd}} = \text{odd}(T)$. Each node is represented as a triplet (u, v, w), where $w = \mathcal{A}_q(u, v)$. The input to the optimization is represented as a set of triplets T which includes the target fundamentals T_{odd} as well as the NOFs, i.e., the complete MCM solution. The maximum number of pipeline stages is initialized to the minimum possible adder depth of the adder graph:

$$S := D_{\text{max}} = \max_{t \in T} \text{AD}_{\text{min}}(t) \tag{3.2}$$

The reason for setting the pipeline depth to the minimum possible is that each additional pipeline stage introduces additional registered operations. So it is unlikely to happen that a better solution with larger pipeline depth exists (this topic is further evaluated in Chapter 4.5.5).

To give an example, the adder graph shown in Figure 3.1(a) has the output fundamentals $T_{\text{odd}} = \{1, 15, 255, 423\}$ and the NOF 47, This leads to a triplet set of $T = \{(1, 1, 1), (1, 1, 15), (1, 15, 47), (1, 1, 255), (47, 47, 423)\}$.

The complete ILP formulation is given in ILP Formulation 1. The first constraint (C1) simply defines that the input node '1' is scheduled in stage zero. Constraint C2 ensures that all output fundamentals are available in the last stage S. Constraint C3 prevents that a node is scheduled before it can be computed. The last constraint (C4) is the most important one which ensures that if w is scheduled in stage s ($x_w^s = 1$), either the corresponding input values u and v to compute w have to be available in the stage before ($x_u^{s-1} = 1$ and $x_v^{s-1} = 1$) or w has to be already computed in the stage before ($x_w^{s-1} = 1$).

The obtained scheduling can be represented as a $|T| \times S$ matrix, where each column represents the pipeline stages in which a node is scheduled. The scheduling matrix for the optimal solution as shown in Figure 3.1(c) using the column order $1, 15, 47, 255, 423$ is equal to

$$\mathbf{X} = \begin{pmatrix} 1 & 1 & 0 & 0 & 0 \\ 1 & 0 & 1 & 0 & 0 \\ 1 & 1 & 0 & 1 & 1 \end{pmatrix}, \tag{3.3}$$

e. g., node '1' is scheduled in all stages whereas node '15' is only scheduled in stage 1 and 3, etc.

ILP Formulation 1 (Adder Graph Pipelining).

$$\min \sum_{s=1}^{S} \sum_{(u,v,w)\in \mathcal{T}} x_w^s$$

subject to

C1: $\quad x_1^0 = 1$

C2: $\quad x_w^S = 1$ for all $w \in T_{\mathrm{odd}}$

C3: $\quad x_w^s = 0$ for all $s < \mathrm{AD}_{\min}(w); (u,v,w) \in \mathcal{T}$

C4: $2x_w^s - 2x_w^{s-1} - x_u^{s-1} - x_v^{s-1} \leq 0$ for all $(u,v,w) \in \mathcal{T}$

$$\text{with } s = \mathrm{AD}_{\min}(w) \ldots S$$

$$x_w^s \in \{0,1\}$$

3.2.2 Extended ILP Formulation

The ILP formulation above can be simply extended to different costs for adders (cost_A) and registers (cost_R). For that, each binary variable x is replaced by two binary variables which represent their realization type:

$$a_w^s = \begin{cases} 1 & \text{when node } w \text{ is computed using an adder in stage } s \\ 0 & \text{otherwise} \end{cases} \tag{3.4}$$

$$r_w^s = \begin{cases} 1 & \text{when node } w \text{ is realized using an register in stage } s \\ 0 & \text{otherwise} \end{cases} \tag{3.5}$$

Now, the ILP formulation can be extended as given in ILP Formulation 2. Constraints C1-C3 are equivalent to C1-C3 of ILP Formulation 1, but extended to two binary variables per node and stage. Constraint C4 ensures that if a register for w is placed in stage s, either another register holding the result of w or an adder computing w needs to be realized in the previous stage ($s-1$). If the node has to be computed in stage s, the existence of the two inputs u and v is constrained by C5.

ILP Formulation 2 (Adder Graph Pipelining with Separate Costs).

$$\min \sum_{s=1}^{S} \sum_{(u,v,w)\in\mathcal{T}} \mathrm{cost_A}(w)a_w^s + \mathrm{cost_R}(w)r_w^s$$

subject to

C1: $r_1^0 = 1$

C2: $a_w^S + r_w^S = 1$ for all $w \in \mathcal{T}_{\mathrm{odd}}$

C3a: $a_w^s = 0$ for all $s < \mathrm{AD_{min}}(w); (u,v,w) \in \mathcal{T}$

C3b: $r_w^s = 0$ for all $s \leq \mathrm{AD_{min}}(w); (u,v,w) \in \mathcal{T}$

C4: $r_w^s - a_w^{s-1} - r_w^{s-1} = 0$ for all $(u,v,w) \in \mathcal{T}$

              with $s = \mathrm{AD_{min}}(w) + 1 \dots S$

C5: $\left.\begin{aligned} a_w^s - r_u^{s-1} - a_u^{s-1} \leq 0 \\ a_w^s - r_v^{s-1} - a_v^{s-1} \leq 0 \end{aligned}\right\}$ for all $(u,v,w) \in \mathcal{T}$ with $s = \mathrm{AD_{min}}(w) \dots S$

$$a_w^s, r_w^s \in \{0,1\}$$

The extended ILP formulation also allows to chose a trade-off between latency and throughput, i. e., placing registers only every n'th stage to decrease the latency at the cost of a lower throughput or using pipelined adders to further increase the throughput at the cost of a higher latency. This can be simply done by using different cost functions for each stage and setting $\mathrm{cost_R}(w) = 0$ for stages which are skipped by wires or increasing the adder/register costs in case of using pipelined adders. However, pipelining every stage typically leads to the best resource/throughput ratio.

3.3 Experimental Results

To evaluate the performance of the proposed optimal BILP pipelining, it is compared to the cut-set retiming method using an ASAP scheduling. First, the overhead in terms of registered operations is considered. Then, synthesis experiments are performed using a VHDL code generator.

The adder graphs in both experiments were obtained by using $\mathrm{H_{cub}}$ [41] which is one of the leading MCM algorithms. Its implementation is available as open-source [104] and a C++ wrapper for Matlab was written using the

Matlab MEX interface [126]. The VHDL code generator produces complete FIR filters and was written in Matlab.

The H_{cub} implementation [104] provides an option to solve the MCM problem with bounded adder depth (see Definition 5 on page 19), denoted as $H_{cub,BAD}$. This typically leads to an increase of adders but also to a minimum pipeline depth. Thus, the impact of the adder depth can now be analyzed. Note that the proposed method can be applied to adder graphs obtained by *any* MCM method.

A set of benchmark filters is analyzed which was already used in the publication of Mirzaei et al. [5]. It contains FIR filters from $N = 6$ to 151 taps. The coefficient values are available online in the FIR benchmark suite [127] labeled as MIRZAEI10_N. Matlab was used to call H_{cub} and to generate an ILP model file, which was then solved by using a standard BILP solver. First, the Matlab BILP solver (`bintprog` method) was used leading to optimization times of up to 36 minutes on an Intel Core i5 CPU with 2.4 GHz. Switching the solver to SCIP [128, 129] reduced the optimization time to less than one tenth of a second on the same machine.

3.3.1 Register Overhead

The optimization results in terms of number of registered operations (adder+register or pure register) for both MCM algorithms (H_{cub} and $H_{cub,BAD}$) and the high-level cost model (i. e., $\text{cost}_A(w) := \text{cost}_R(w) := 1$) are listed in Table 3.1. ILP Formulations 1 and 2 were both used and lead to identical objective values as expected. However, ILP Formulation 1 was slightly faster due to the reduced variables and constraints. It can be observed that H_{cub} always finds an adder graph with equal or less adders compared to $H_{cub,BAD}$ (6.5% less on average). However, this is obtained at the expense of an increased AD which ranges from 3 to 9 for H_{cub} compared to an AD of 3 for $H_{cub,BAD}$. The proposed optimal pipelining method leads to a reduction of registered operations in most of the cases compared to the conventional CSR with ASAP schedule. The percentage improvement is listed in column 'imp.'. Only for two small instances ($N = 10$ and $N = 13$ for $H_{cub,BAD}$) the CSR method leads to the same result. On average, the improvement is 24.4% and 5.8% for H_{cub} and $H_{cub,BAD}$, respectively. Clearly, the best results in terms of registered operations are achieved for adder graphs with low AD. Although a higher number of adders is used by $H_{cub,BAD}$ this is compensated by the reduced registered operations which are 19.4% less on average compared to H_{cub} using the proposed optimal

Table 3.1: Optimization results for H_{cub} and $H_{cub,BAD}$ using the benchmark set of [5], without pipelining (no. of adders), CSR pipelining with ASAP scheduling and the proposed optimal pipelining (registered operations).

| N | H_{cub} | | | | | $H_{cub,BAD}$ | | | | |
| | AD | no. of adders | reg. operations | | | AD | no. of adders | reg. operations | | |
			CSR	opt. pip.	imp.			CSR	opt. pip.	imp.
6	3	6	11	10	9.1%	3	6	11	10	9.1%
10	4	9	19	17	10.5%	3	9	14	14	0%
13	5	11	30	25	16.7%	3	13	20	20	0%
20	9	11	56	36	35.7%	3	14	22	21	4.5%
28	6	17	47	36	23.4%	3	20	29	26	10.3%
41	6	22	91	50	45.1%	3	24	39	38	2.6%
61	6	32	108	69	36.1%	3	35	52	47	9.6%
119	3	53	89	76	14.6%	3	53	89	76	14.6%
151	4	70	150	107	28.7%	3	73	92	91	1.1%
avg.:	5.11	25.7	66.8	47.3	24.4%	3.0	27.4	40.9	38.11	5.8%

pipelining.

The pipelined graphs were analyzed for the usage of adder duplications. A total number of 21 operations could be saved by using adder duplication for adder graphs obtained by H_{cub} but only two operations could be saved for adder graphs obtained by $H_{cub,BAD}$. It can be concluded that the choice of the MCM algorithm and the scheduling/pipelining method significantly affects the total number of pipelined operations. Furthermore, MCM algorithms for reduced adder depth are highly recommendable for pipelined adder graphs.

3.3.2 Slice Comparison

So far, only the number of registered operations (which include registered adders as well as pure registers) were considered as cost metric to approximate the real complexity of pipelined adder graphs. Hence, synthesis experiments were performed to obtain the real FPGA resource consumption as well as the speed of the circuit. For that, a VHDL code generator was developed for FIR filters using the adder graphs (with or without pipelining) for constant multiplications. The optimal pipelining was applied to adder graphs obtained by $H_{cub,BAD}$. Each filter is compared with three other methods:

- Pipelined FIR filter optimization method 'add and shift' of Mirzaei et al. [5, 6]

- H_{cub} without pipelining [41]

- DSP-based design using Xilinx Coregen FIR Compiler (v4.0) [130]

All designs were synthesized for a Xilinx Virtex 4 FPGA (XC4VLX100-10 and XC4VSX55-10 for DSP-based filters due to their high DSP count) like done by Mirzaei et al. [6] using the Xilinx Integrated Software Environment (ISE) 11.1 and were optimized for speed. Synthesis results from the add/shift method were directly taken from the publication [6]. Note that the corresponding Verilog implementations which are provided online at [131] contain incorrect pipelines (some registers are missing to balance the pipeline). However, it is assumed that this error was not present in the original work, otherwise it would consume more resources when adding the missing registers. The inputs and outputs of the filters are registered for a fair speed comparison. The input word size has been set to 12 bit as in [6]. All filters were implemented with full precision without any truncation.

The numeric results are listed in Table 3.2 and illustrated in Figure 3.2. Compared to the add/shift method, the proposed combination of $H_{cub,BAD}$ with the optimal pipelining leads to a slice improvement of 54.1% on average while a similar performance (11.8% decrease in clock frequency) is obtained. As expected, the direct implementation of adder graphs obtained by H_{cub} (without pipelining) leads to a much lower speed of about 100 MHz which is far below the speed capabilities of the FPGA. The speed improvement of the proposed method compared to H_{cub} is 308.0% on average. However, the speed improvement comes at a price. The slice resources increased by 17.9% due to additional pipeline registers. However, this overhead is worthwhile with respect to the achieved speedup.

3.3.3 Device Utilization

Besides the adder graph based solutions discussed so far, Table 3.2 also provides results when DSP blocks are used (column "DSP-based"). A fair comparison of the resources between slice dominated designs and DSP block designs is very difficult, if not impossible. In other words, it is hard to answer the question if a design with 1000 slices is better than a design with 10 DSP blocks as both designs may be optimal in a Pareto sense. However, to

Table 3.2: Synthesis results of benchmark filters from [6] on Xilinx Virtex 4 for the proposed method (optimal pipelining using $H_{cub,BAD}$) compared to other methods discussed. N is the number of coefficients

	Add/shift [6]		H_{cub}		DSP-based				proposed	
N	Slices	f_{max} [MHz]	Slices	f_{max} [MHz]	Slices	DSP48	Slice Equiv.	f_{max} [MHz]	Slices	f_{max} [MHz]
6	264	296	131	115.5	36	6	144	402.7	176	320.9
10	475	296	227	100.4	44	10	224	402.1	311	273.2
13	387	296	280	80.5	41	13	275	406.3	400	275.1
20	851	271	396	53.9	57	20	417	405.2	527	267.3
28	1303	305	595	67.7	60	28	564	401.8	704	250.2
41	2178	296	779	73.5	70	41	808	400.6	1019	230.0
61	3284	247	1185	69.1	76	61	1174	400.8	1397	248.8
119	6025	294	2243	96.4	370	119	2512	372.7	2609	214.8
151	7623	294	2880	85.9	612	151	3330	351.4	3131	208.3
avg.:	2487.8	288.3	968.4	82.5	151.8	49.9	1049.8	393.7	1141.6	254.3

compare two Pareto points, a weighted sum is often used in single-objective optimization methods when applied to multi-objective problems. The problem is still how to choose the weights. One solution is to weight each resource with the inverse of the amount of available resources on a specific device divided by the number of the different resources. This measure can intuitively be seen as a *device utilization* as it results in 100% if every resource is completely used.

This metric was applied to each device of the Virtex 4 device family using the average slice and DSP block utilizations of the proposed method (1142 slices) and the MAC-based realization (152 slices and 50 DSP48). The results are shown in Table 3.3. It can be seen that the slice per DSP48 ratio of the Virtex 4 device family highly varies from 48 to 928. Even for the lowest slice/DSP48 ratio, the MAC-based realization consumes 5.2% of the FPGA in total (average of 0.6% slices and 9.8% DSP48) whereas the proposed method requires only 2.3% (4.6% slices, 0% DSP48). For the highest slice/DSP48 ratio, the MAC realization requires 26.2% (0.2% slice, 52.1% DSP48) compared to 0.6% (1.3% slice, 0% DSP48) for the proposed method. Note that only four out of 17 FPGAs provide the DSP resources for implementing the largest filter ($N = 151$) in the benchmark set.

This device utilization metric shows that the multiplication with (multiple) constants using slice resources is a valuable method when DSP blocks are limited. Note that the slice/DSP ratio is decreasing with newer FPGA

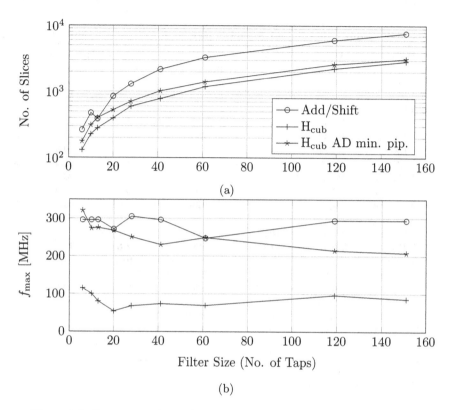

Figure 3.2: Resource and speed comparison of adder graph based methods

families. The lowest slice/DSP ratios for Virtex 6 and Virtex 7 are 36.6 and 27.1, respectively, compared to the lowest slice/DSP ratio of 48 of Virtex 4.

3.3.4 Silicon Area

While the device utilization as defined in the last section highlights the user's point-of-view, the occupied chip area is of general interest. It can at least answer the question if a DSP block is efficient for constant multiplications or if it is more economic to take a probably cheaper FPGA with larger slice/DSP ratio. Unfortunately, the silicon area of the building blocks of commercial FPGAs is generally not accessible. However, the silicon area of a slice and the embedded multiplier of a Virtex II 1000 was obtained by

Table 3.3: Device utilization using a mean metric for the Virtex 4 device family

Device	Avail. Slices	Avail. DSP48	Avail. Slice/DSP48	Device utilization (%)	
(XC4...)				proposed	MAC
VLX15	6144	32	192	9.3	79.4
VLX25	10752	48	224	5.3	52.8
VLX40	18432	64	288	3.1	39.5
VLX60	26624	64	416	2.1	39.3
VLX80	35840	80	448	1.6	31.5
VLX100	49152	96	512	1.2	26.2
VLX160	67584	96	704	0.8	26.2
VLX200	**89088**	**96**	**928**	**0.6**	**26.1**
VSX25	10240	128	80	5.6	20.3
VSX35	15360	192	80	3.7	13.5
VSX55	**24576**	**512**	**48**	**2.3**	**5.2**
VFX12	5472	32	171	10.4	79.5
VFX20	8544	32	267	6.7	79.0
VFX40	18624	48	388	3.1	52.5
VFX60	25280	128	197.5	2.3	19.8
VFX100	42176	160	263.6	1.4	15.8
VFX140	63168	192	329	0.9	13.1
avg.:	30415	117.6	325.7	3.6	36.5

die photo inspections in previous work [132]. The complexity of a Virtex II CLB is similar to that of the Virtex 4 architecture, as both contain four slices where each slice contains two 4-input LUTs, two FFs and carry-chain resources [59, 133]. The 18×18 bit multiplier of Virtex II was extended to the DSP48 block in the Virtex 4 architecture by adding a post-adder and additional routing resources [133, 134]. This extension complicates a direct transfer of the area requirement to the Virtex 4 architecture, but can serve as a lower bound. The multiplier to slice ratio was reported to be 1:18 [132], i. e., one embedded multiplier requires an equivalent chip area of 18 slices. This ratio was used to compute a lower bound of a *slice equivalent*, which is computed by the number of DSP blocks times 18 plus the number of slices. The slice equivalent is also listed for the DSP-based designs in Table 3.2. It shows that the equivalent slices using DSP blocks are in the same order of magnitude than logic-based MCM designs reaching a similar or even higher complexity for large N.

3.4 Conclusion

It was shown in this chapter that pipelining of adder graphs for multiple constant multiplication has to be done very carefully. First, the optimization method to obtain the adder graph should consider solutions with minimal adder depth. Second, the scheduling of operations has to be optimized to yield the least overhead in additional pipeline registers. Two ILP formulations were proposed to achieve the optimal schedule, one simple formulation using a single cost function per registered operation (useful for FPGAs with single FFs per BLE) and another formulation with separate cost functions for adders and registers (useful for FPGAs with multiple FFs per BLE or ASICs).

The proposed optimization leads to filter designs that have a significant lower resource usage compared to the 'add/shift' method proposed by Mirzaei et al. [5, 6] while having a similar performance. As it was shown that the 'add/shift' method needs less resources than parallel distributed arithmetic (DA) implementations [5, 6, 118], it can be concluded that the proposed method also outperforms parallel DA implementations.

The use of adder graphs for constant multiplications leads to a reduced device utilization compared to DSP-based implementations in the Virtex 4 device family. Only a few slices are needed on average for each constant multiplication. This makes the method attractive for small FPGAs with low embedded multiplier count or applications where generic multipliers are needed for other tasks. However, as demonstrated in the next two chapters, it is usually much better to consider the register cost of a pipelined MCM block directly in the optimization. The results from this chapter will serve as a bridge to be able to compare pipelined MCM solutions with conventional MCM algorithms which are pipelined afterwards.

4 The Reduced Pipelined Adder Graph Algorithm

It was shown in the last chapter that the quality of a pipelined adder graph (PAG) strongly depends on the underlying adder graph as well as the pipelining method. This fact is illustrated using Figure 4.1 showing different MCM realization for the coefficient set $\{44, 130, 172\}$. Figure 4.1(a) shows an optimal adder graph (in the sense of the adder count) which was obtained from the optimal MCM algorithm of Aksoy et al. [99]. The PAG of this adder graph using the optimal pipelining of the previous chapter is shown in Figure 4.1(b). It uses seven registered operations which are distributed over three stages. Limiting the adder depth leads to an increase of adders in the adder graph but also to a reduction of registered operations in the PAG as shown in figures 4.1(c) and 4.1(d), respectively. Now, only six registered operations are needed. However, as demonstrated in Figure 4.1(e), there exists a PAG that can compute the same operation with only five registered operations. It is evident that even for such small problems, there is a big potential to decrease resources compared to the optimal MCM solution together with the optimal pipelining method of the last chapter. Hence, a direct optimization of pipelined adder graphs is considered in this chapter. First, the optimization problem, called pipelined MCM (PMCM) problem, is formally defined. Then, a heuristic method, the reduced pipelined adder graph (RPAG) algorithm, is presented. The work presented in this chapter is an extension of the work originally published in [8]. The extensions can be summarized as follows:

- Several depth first search iterations instead of a single greedy search are performed

- Introduction of a new topology (see Figure 4.2(d))

- Modified selection criteria (gain function)

Its implementation is freely available as open-source [16].

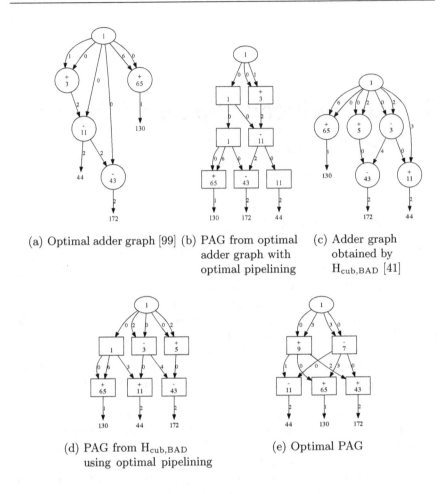

(a) Optimal adder graph [99] (b) PAG from optimal   (c) Adder graph
                              adder graph with   obtained by
                              optimal pipelining $H_{cub,BAD}$ [41]

(d) PAG from $H_{cub,BAD}$ (e) Optimal PAG
    using optimal pipelining

Figure 4.1: Different multiplier block realizations for the coefficients $\{44, 130, 172\}$

4.1 Definition of the PMCM Problem

As discussed in Section 2.4.4, the main effort during MCM optimization is
the determination of the non-output fundamentals (NOFs) [41], i.e., the
nodes which are needed to compute the output nodes (e.g., node '3' in Figure 4.1(a)). Once all NOF values are known it is a simple combinational
task to compute the necessary edge connections and shifts by using the optimal part of common MCM algorithms (see Algorithm 2.1 on page 31). This

simplifies the implementation of the MCM optimization as the shift factors, signs and node dependencies do not have to be stored during runtime. For the example set above, it is sufficient to know that the NOF 3 has to be added to the target set $\{44, 130, 172\}$ to be able to compute all values from input node 1. This principle can be adapted to the optimization of PAGs. For that, each NOF has to be provided with a stage information. Thus, if it is known that the output stage of the example in Figure 4.1(e) can be computed from $\{7, 9\}$ in stage 1, the necessary edges and shift values can be simply computed in the same way as for MCM. Hence, it is appropriate to define sets X_s containing the node values for each pipeline stage s. The pipeline sets for the PAG in Figure 4.1(e) are, e.g., $X_0 = \{1\}$, $X_1 = \{7, 9\}$ and $X_2 = \{11, 43, 65\}$. The core problem now is to find the minimal sets X_s for the intermediate stages $s = 1 \dots S - 1$ where S is the number of pipeline stages in total. This can be formally defined as follows:

Definition 7 (Pipelined MCM Problem). *Given a set of positive target constants* $T = \{t_1, \dots, t_M\}$ *and the number of pipeline stages* S, *find sets* $X_1, \dots, X_{S-1} \subseteq \{1, 3, \dots, c_{\max}\}$ *with minimal area cost such that for all* $w \in X_s$ *for* $s = 1, \dots, S$ *there exists a valid* $\mathcal{A}$-*configuration* q *such that* $w = \mathcal{A}_q(u, v)$ *with* $u, v \in X_{s-1}$, $X_0 = \{1\}$ *and* $X_S = \{\mathrm{odd}(t) \mid t \in T \setminus \{0\}\}$.

The *area cost* of the pipeline sets $X_1, \dots, X_{S-1}$ depends on the optimization target and will be discussed in the following.

4.1.1 Cost Definition

The simplest cost metric for evaluating the performance of a PAG is counting the number of nodes where each node represents a *registered operation*. This cost model was introduced as the high-level area cost model in Section 2.5.4 and can be directly obtained from the pipeline sets:

$$\mathrm{cost}_{\mathrm{PAG,HL}} = \sum_{s=1}^{S} |X_s| \tag{4.1}$$

This metric is a good estimate for FPGAs as registers have a similar resource consumption as registered adders.

However, the cost at BLE level might differ as introduced in Section 2.5.4. To represent that, each pipeline set X_s is divided into a set of pure registers $R_s = X_s \cap X_{s-1}$ (elements that are repeated from one stage to the next) and a set of registered adders $A_s = X_s \setminus X_{s-1}$ (elements that have to be computed in stage s). Note that their union results in the pipeline sets $X_s = R_s \cup A_s$. Then, the low-level area cost model of a PAG is

$$\text{cost}_{\text{PAG,LL}} = \sum_{s=1}^{S} \left(\sum_{a \in A_s} \text{cost}_A(a) + \sum_{r \in R_s} \text{cost}_R(r) \right) , \qquad (4.2)$$

with the cost functions as defined in Section 2.5.4.

4.1.2 Relation of PMCM to MCM

The PMCM problem as defined above can be seen as a generalization of the MCM problem as the MCM problem can be solved by using a PMCM solver by setting the register cost to zero ($\text{cost}_R(c) = 0$) and the pipeline depth S to a sufficiently large number. Of course, as the search space significantly grows with increasing S, it is not very attractive to do so for generic MCM problems. However, the PMCM problem can be easily adopted to MCM problems with adder depth constraints (see Section 2.4.5). For the MCM_{BAD} problem (see Definition 5 on page 19), the register costs are simply set to zero while S remains to be the minimum possible like for PMCM. For the MCM_{MAD} problem (see Definition 6 on page 19), the pipeline sets additionally have to be initialized in a different way. While for PMCM, the pipeline sets are initialized to

$$X_0 := \{1\} \qquad (4.3)$$
$$X_S := \{\text{odd}(t) \mid t \in T\} \setminus \{0\} \qquad (4.4)$$

with $X_s := \emptyset$ for $s = 1 \ldots S - 1$, they have to be initialized to

$$X_s := \{\text{odd}(t) \mid \text{AD}_{\min}(t) = s, \ t \in T\} \text{ for } s = 1 \ldots S \qquad (4.5)$$

for the MCM_{MAD} problem. For the example above with the odd target set $\{11, 65, 43\}$, the pipeline sets for the MCM_{MAD} problem would be initialized to $X_0 = \{1\}$, $X_1 = \{65\}$, $X_2 = \{11\}$ and $X_3 = \{43\}$ as $\text{AD}_{\min}(65) = 1$, $\text{AD}_{\min}(11) = 2$ and $\text{AD}_{\min}(43) = 3$. Hence, any method solving the PMCM problem as defined above can be easily adapted to the MCM_{BAD} or MCM_{MAD} problem. This will be demonstrated in Section 5.6.6.

4.2 The RPAG Algorithm

An algorithm to solve the PMCM problem as well as the MCM problems with adder depth constraints is presented in the following. It is called reduced pipelined adder graph (RPAG) algorithm and is divided into two parts: The inner loop follows a depth-first search (DFS) approach and is a modified version of the original greedy search algorithm presented in [8]. The outer loop controls the direction of the DFS in a simple way. The strategy is to perform a greedy search in the first depth search iteration which is then extended to solutions which lie in the neighborhood of this initial greedy solution. In the following, the RPAG ideas are presented by a basic algorithm for PMCM according to [8] which corresponds to the inner loop. It is later extended by the outer loop to improve its optimization quality and MCM problems with adder depth constraints.

4.2.1 The Basic RPAG Algorithm

In contrast to most existing MCM algorithms, the optimization in the basic RPAG algorithm starts form the output nodes of the adder graph and searches for preceding nodes until the input node '1' is reached. Hence, starting with stage $s = S$, the elements in stage $s - 1$ are searched until all elements in X_s can be computed from X_{s-1}. Then, the next lower pipeline stage is optimized. The pseudo code of the basic RPAG algorithm is shown in Algorithm 4.1 and is described in the following:

1. Given a set of target coefficients T, the pipeline depth S is set to the maximum of the minimum adder depths of all the elements in target set T (line 2); pipeline set X_S is initialized to the odd representation of the target coefficients (line 3).

2. For each stage, the elements of the current stage are copied into a working set $W = X_s$ and the elements of the preceding stage, called predecessors, are initialized to an empty predecessor set P (lines 5 and 6).

3. Next, valid predecessors are searched and evaluated for all remaining elements in W. The search is started by computing all possible single predecessors and the best one, if exists, is returned (function best_single_predecessor(), line 8). If a single predecessor was found, it is added in P (line 10) and all elements that can be realized from this predecessor are removed from W (line 15).

4. If no single predecessor was found (indicated by $p = 0$), a pair of predecessors (best_predecessor_pair(), line 12) is searched. A valid pair can always be found and the best pair found is inserted into P (line 13); the corresponding elements that can now be computed from

Algorithm 4.1: The basic RPAG algorithm

1 RPAG (T)
2    $S := \max_{t \in T} \text{AD}_{\min}(t)$
3    $X_S := \{\text{odd}(t) \mid t \in T\} \setminus \{0\}$
4   for $s = S \ldots 2$
5      $W := X_s$
6      $P := \emptyset$
7   do
8        $p \leftarrow \text{best\_single\_predecessor}(P, W, s)$
9   if $p \neq 0$
10          $P \leftarrow P \cup \{p\}$
11   else
12          $(p_1, p_2) \leftarrow \text{best\_predecessor\_pair}(W, s)$
13          $P \leftarrow P \cup \{p_1, p_2\}$
14   end if
15        $W \leftarrow W \setminus \mathcal{A}_*(P)$
16   while $|W| \neq 0$
17      $X_{s-1} \leftarrow P$
18   end for

P are removed from W (line 15).

5. Steps 3 and 4 are repeated until the working set is empty, meaning that all elements in X_s can be computed from P. Next, P is copied into X_{s-1} (line 17) and s is decremented. The algorithm terminates when stage $s = 2$ was computed, i.e., X_1 is determined.

The most important step is the determination of predecessors which is presented in the following.

4.2.2 Computation of Single Predecessors

The computation of all single predecessors (best_single_predecessor() in Algorithm 4.1) is done by evaluating the graph topologies shown in Figure 4.2 (a)-(c). Each predecessor p must have a lower adder depth than the current stage, i.e.,

$$\text{AD}_{\min}(p) < s . \tag{4.6}$$

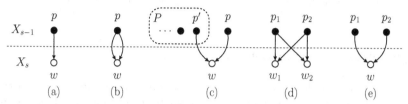

Figure 4.2: Predecessor graph topologies for (a)-(c) a single predecessor p, (d)-(e) a predecessor pair (p_1, p_2)

Topology (a) occurs if W contains elements with a lower adder depth than s. These elements can be copied to P and are realized using pure registers (for PMCM) or wires (for MCM). In topology (b), an element of W is computed from a single predecessor by multiplying the predecessor with a number of the form $2^k \pm 1$ ($k \in \mathbb{N}_0$). Hence, valid predecessors of that topology are all integers obtained by the division of the working set vectors by $2^k \pm 1$. If P already contains some elements, more elements can be computed from a single predecessor using one $\mathcal{A}$-operation as shown in Figure 4.2 (c). To obtain the corresponding predecessor p, Lemma 2 (see page 16) can be used. Hence, predecessors corresponding to that topology can be computed from $p = \mathcal{A}_q(w, p')$ for $w \in W$ and $p' \in P$ with $\mathrm{AD_{min}}(p) < s$. The computations can now be formally represented by the sets:

$$P_a = \{w \in W \mid \mathrm{AD_{min}}(w) < s\} \tag{4.7}$$

$$P_b = \{w/(2^k \pm 1) \mid w \in W, \, k \in \mathbb{N}\} \cap \mathbb{N} \setminus \{w\} \tag{4.8}$$

$$P_c = \bigcup_{\substack{w \in W \\ p' \in P}} \{p = \mathcal{A}_q(w, p') \mid q \text{ valid}, \, \mathrm{AD_{min}}(p) < s\} \tag{4.9}$$

4.2.3 Computation of Predecessor Pairs

If none of the working set elements can be computed from a single predecessor, a pair of predecessors is searched (`best_predecessor_pair()` in Algorithm 4.1). However, the enumeration of all possible predecessor pairs is very time consuming as *any* integer p_1 with lower adder depth has to be evaluated and for each of them there will be typically several valid p_2 elements to compute w. Hence, the search space has to be pruned to good candidates to make the search feasible. This is performed by evaluating the topologies shown in figures 4.2 (d) and 4.2 (e) which are described in the following.

High Potential Predecessor Pairs

If there exist a predecessor pair from which two or more elements from the working set can be computed, it is covered by the graph topology (d) as shown in Figure 4.2. There exist other two-predecessor topologies including multiplier structures like topology (b) but the corresponding predecessors were already selected in the single predecessor search.

Both elements w_1 and w_2 are per definition a result of the $\mathcal{A}$-operations

$$w_1 = \mathcal{A}_{q1}(p_1, p_2)$$
$$w_2 = \mathcal{A}_{q2}(p_1, p_2) \tag{4.10}$$

which can be rewritten with extra sign bits to avoid a case discrimination when evaluating the $\mathcal{A}$-operation (2.6):

$$w_1 = (-1)^{s_{11}} \left(2^{l_{11}} p_1 + (-1)^{s_{21}} 2^{l_{21}} p_2 \right) 2^{-r_1} \tag{4.11}$$

$$w_2 = (-1)^{s_{12}} \left(2^{l_{12}} p_1 + (-1)^{s_{22}} 2^{l_{22}} p_2 \right) 2^{-r_2} \tag{4.12}$$

This equation system can be solved for p_1 and p_2, leading to

$$p_1 = \frac{(-1)^{s_{11}}(-1)^{s_{21}} w_1 2^{r_1 - l_{21}} - (-1)^{s_{12}}(-1)^{s_{22}} w_2 2^{r_2 - l_{22}}}{(-1)^{s_{21}} 2^{l_{11} - l_{21}} - (-1)^{s_{22}} 2^{l_{12} - l_{22}}} \tag{4.13}$$

$$p_2 = \frac{(-1)^{s_{11}} w_1 2^{r_1 - l_{11}} - (-1)^{s_{12}} w_2 2^{r_2 - l_{12}}}{(-1)^{s_{21}} 2^{l_{21} - l_{11}} - (-1)^{s_{22}} 2^{l_{22} - l_{12}}} . \tag{4.14}$$

Their positive odd fundamental representation yields

$$p_1 = \left| \frac{w_1 2^{r_1 + l_{22}} - (-1)^{s_{12}}(-1)^{s_{22}} w_2 2^{r_2 + l_{21}}}{2^{l_{11} + l_{22}} - (-1)^{s_{22}} 2^{l_{12} + l_{21}}} \right| \tag{4.15}$$

$$p_2 = \left| \frac{w_1 2^{r_1 + l_{12}} - (-1)^{s_{12}} w_2 2^{r_2 + l_{11}}}{2^{l_{21} + l_{12}} - (-1)^{s_{22}} 2^{l_{22} + l_{11}}} \right| \tag{4.16}$$

where each fraction was extended by an appropriate power-of-two value to allow the computation of numerators and denominators with integer arithmetic.

To evaluate all possible odd predecessors, not all variable combinations have to be considered in (4.15) and (4.16). As shown in Appendix B, only the following cases have to be considered for an $\mathcal{A}$-operation:

i) $l_u \geq 1$, $l_v = r = 0$

ii) $l_v \geq 1$, $l_u = r = 0$

iii) $r \geq 0$, $l_u = l_v = 0$

Hence, the evaluation can be reduced to the following combinations:

i) $l_{11} \geq 1$, $l_{12} \geq 1$, $l_{21} = l_{22} = r_1 = r_2 = 0$ (p_1 shifted for w_1, p_1 shifted for w_2)

ii) $l_{11} \geq 1$, $l_{22} \geq 1$, $l_{12} = l_{21} = r_1 = r_2 = 0$ (p_1 shifted for w_1, p_2 shifted for w_2)

iii) $l_{11} \geq 1$, $r_2 \geq 0$, $l_{12} = l_{21} = l_{22} = r_1 = 0$ (p_1 shifted for w_1, result shifted for w_2)

iv) $r_1 \geq 0$, $r_2 \geq 0$, $l_{11} = l_{12} = l_{21} = l_{22} = 0$ (result shifted for w_1 and w_2)

The computation has to be done two times where w_1 and w_2 are swapped in the second run.

It may happen that more than two working set elements can be computed from topology (d). This is considered in the evaluation of the potential predecessors and will be described in Section 4.2.4. Of course, there is no guarantee that any predecessor pair is found by this topology. However, in case of failure, it is known that only one working set element can be reduced by including two predecessors. One could continue to evaluate three or more predecessors but then, the search space becomes huge as the resulting equation system is underdetermined. Instead, an MSD-based evaluation is performed in RPAG which is described next.

MSD-based Predecessor Pairs

In the MSD-based predecessor search, predecessor pairs are obtained by distributing the non-zero digits of the MSD representations of w to the predecessors. This guarantees the construction of at least one element of the working set as shown in Figure 4.2 (e), i. e., it will not fail to find a predecessor like in the other topologies. Although this is still a large search space it turned out to be tractable for practical problems (see Section 4.2.6). The MSD representation also has the advantage that a reduced adder depth can be guaranteed by construction.

The generation of predecessors is explained by using the example coefficient 13. Its MSD representations are $10\bar{1}01$, $100\bar{1}\bar{1}$ and 1101. A predecessor pair (p_1, p_2) can now be extracted by assigning some bits of the MSD representation to p_1 and the other bits to p_2. Of course, the AD of each predecessor p must fulfill (4.6), which is given if the number of non-zeros of each predecessor fulfills

$$\text{nz}(p) \leq 2^{\text{AD}_{\min}(w)-1} \ . \tag{4.17}$$

The number of nonzero digits of coefficient 13 is $\text{nz}(13) = 3$ leading to a minimum AD of two, according to (2.18) (see page 18). To reduce the AD to $s = 1$, each predecessor can have at most $\text{nz}(p) = 2$ bits in its MSD representation, leading to the following predecessor pairs: $(10000, \bar{1}01)$, $(10\bar{1}00, 1)$, $(10001, \bar{1}00)$, $(10000, \bar{1}\bar{1})$, $(1000\bar{1}, \bar{1}0)$, $(100\bar{1}0, \bar{1})$, $(1000, 101)$, $(1100, 1)$, $(1001, 100)$. Taking the odd representation of their absolute values yields the decimal predecessor pairs $(1, 3)$, $(1, 5)$, $(1, 7)$, $(1, 9)$, $(1, 15)$ and $(1, 17)$. This procedure is performed for all MSD representations of all elements in W.

4.2.4 Evaluation of Predecessors

One crucial point in the overall optimization is the local evaluation of predecessors. The basic idea is to prefer the least number of predecessors which are able to compute the most elements from the working set.

The set of predecessors to be evaluated is denoted by P', which is set to $P' = \{p\}$ if a single predecessor is evaluated and is set to $P' = \{p_1, p_2\}$ if a pair of predecessors is evaluated. The set of elements which can be removed from the working set is denoted as

$$W'(P') = W \cap \mathcal{A}_*(P \cup P') \ . \tag{4.18}$$

Then, the basic strategy of RPAG is to select the predecessor(s) which yield to the highest benefit-to-cost ratio, which is called *gain* in the following:

$$\text{gain}(P') = \frac{|W'(P')|}{|P'|} \tag{4.19}$$

So far, this metric does neither respect the real cost nor distinguish between registers or adders. For this, we divide the working set into a set of elements which are realized by registers and another set which are realized by adders:

$$W'_R(P') = W'(P') \cap P' \tag{4.20}$$

$$W'_A(P') = W'(P') \setminus P' \tag{4.21}$$

A similar division is done with the predecessor set P'. As the information on how the elements are computed is not known in the current stage, it is decided from the adder depth if the predecessor has to be computed by an adder or can potentially be realized by a register (if its adder depth is lower than necessary):

$$P'_R(P') = \{p \in P' \mid \mathrm{AD}_{\min}(p) < s - 1\} \tag{4.22}$$

$$P'_A(P') = \{p \in P' \mid \mathrm{AD}_{\min}(p) = s - 1\} \tag{4.23}$$

Now, the benefit-to-cost ratio can be described by

$$\mathrm{gain}(P') = \frac{\sum_{w \in W_R(P')} \mathrm{cost}_R(w) + \sum_{w \in W_A(P')} \mathrm{cost}_A(w)}{\sum_{p \in P'_R(P')} \mathrm{cost}_R(p) + \sum_{p \in P'_A(P')} \mathrm{cost}_A(p)}, \tag{4.24}$$

with the cost functions as defined in Section 2.5.4. Now, costly elements that are computed from cheap predecessors produce the highest gain. This metric still prefers predecessors that are able to produce the most elements in W but also respects their realization cost. Note that this gain definition is a generalization of (4.19) as it is identical in case of $\mathrm{cost}_A(x) = \mathrm{cost}_R(x) = 1$ as $W'_R(P') \cup W'_A(P') = W'(P')$ and $P'_R(P') \cup P'_A(P') = P'$. If the register costs are zero (for MCM problems), the denominator is checked for zero and the gain is set to a sufficiently large number in this case.

The predecessor set P' with highest gain is selected. In case that two or more sets P' result in the same gain, the set with the least average magnitude is chosen. This simple second selection criteria showed better average performance in a large benchmark compared to selecting the predecessors with lowest non-zero count.

4.2.5 The Overall RPAG Algorithm

The evaluation and selection of elements in the basic algorithm above can only be done from a local point of view, i. e., elements which are good for the current stage may be difficult for the optimization of earlier stages (or later optimization steps). Hence, several depth search iterations are performed in the overall algorithm, which is denoted as RPAG (v2) in the following to distinguish between the original RPAG algorithm [8], denoted as RPAG (v1), which corresponds to the basic algorithm above. The first search corresponds to a pure greedy search as described above, i. e., the best element is selected from a local evaluation. In the next search, the first predecessor selection is forced to the second best gain while the remaining elements are still selected in a greedy manner. This is continued up to a certain limit of L additional selections, such that all of the $L + 1$ best predecessors in the first selection are evaluated. Now, the first selection which yields the globally best solution is fixed and the same procedure is continued for the following selections until all selections are fixed. This can be interpreted as a greedy search (overall algorithm/outer loop) on top of another greedy search (depth search/inner loop).

The algorithmic realization is done by the introduction of a *decision vector* $\vec{d}$ which specifies the *decision distance* from the locally best selection. In the first run, the decision vector is not initialized and each time a predecessor selection is *decided*, a zero element is added to $\vec{d}$, meaning that the locally best predecessor was selected (pure greedy selection). After the first run, the number of elements in $\vec{d}$ corresponds to the number of greedy decisions taken during the first iteration. Now, the decision vector can be modified and passed to the next iteration. Assume that four decisions were made in the first (greedy) run, passing, e.g., a decision vector $\vec{d} = (0, 1, 3, 3)^T$ means that the locally best predecessor is selected in the first decision (as $\vec{d}(0) = 0$), the second best predecessor is selected in the second decision (as $\vec{d}(1) = 1$) and the fourth best predecessor is selected in decisions three and four (as $\vec{d}(2) = \vec{d}(3) = 3$). During the first outer loop iterations, the decision vector is modified from $\vec{d} = (0, 0, 0, 0)^T$ to $\vec{d} = (L, 0, 0, 0)^T$ and the globally best solution is saved. Then, d_{best} corresponds to the best decision and is fixed. Next, the second decision is modified by evaluating $\vec{d} = (d_{\text{best}}, 0, 0, 0)^T$ to $\vec{d} = (d_{\text{best}}, L, 0, 0)^T$. This is continued until all decisions are evaluated within the search width limit L. If less predecessors than L are available in the current decision, the next decision is directly selected. Note that the length of the decision vector may vary during the selection.

The complete RPAG (v2) algorithm including its extension to MCM with adder depth constraints is given in Algorithm 4.2. In the first part (lines 2-13), the number of pipeline stages S as well as the pipeline sets $X_{1...S}$ are initialized. This is done in dependence of the specified goal. The odd target fundamentals are all located in stage S for the goal PMCM or MCM_BAD (line 8). In case the goal is set to MCM_MAD, the odd target fundamentals are distributed according to their adder depth in lines 10-12 (see Section 4.1.2). In case of MCM_BAD or MCM_MAD, the register cost are set to zero (not shown in Algorithm 4.2). The cost of the best global PAG solution cost$_{\text{PAG,best}}$, the decision index k and the decision vector $\vec{d}$ are initialized in lines 15-17. Now, in each outer loop (lines 19 to 37) one RPAG depth search (RPAG_DS)

Algorithm 4.2: The RPAG (v2) algorithm

```
1   RPAG(T,L,goal)
```

2    $T_{\mathrm{odd}} := \{\mathrm{odd}(t) \mid t \in T\} \setminus \{0\}$

3    $S := \max_{t \in T_{\mathrm{odd}}} \mathrm{AD}_{\min}(t)$

4    $X_0 := \{1\}$

5    $X_{1\ldots S-1} = \emptyset$

6

```
7    if goal ≡ PMCM or goal ≡ MCM_BAD
```

8      $X_S := T_{\mathrm{odd}}$

```
9    else if goal ≡ MCM_MAD
10       for  t ∈ T_odd
```

11        $X_{\mathrm{AD}_{\min}(t)} \leftarrow X_{\mathrm{AD}_{\min}(t)} \cup \{t\}$

```
12       end for
13   end if
14
```

15    $\mathrm{cost}_{\mathrm{PAG,best}} := \infty$

16    $k := 0$

17    $\vec{d} := \emptyset$

18

```
19   loop
```

20      $X'_{1\ldots S}, \vec{d} \leftarrow$ `RPAG_DS(`$X_{1\ldots S}, \vec{d}, S$`)`

21

22   `if` $\mathrm{cost}_{\mathrm{PAG}}(X'_{1\ldots S}) < \mathrm{cost}_{\mathrm{PAG,best}}$

23        $\mathrm{cost}_{\mathrm{PAG,best}} \leftarrow \mathrm{cost}_{\mathrm{PAG}}(X'_{1\ldots S})$

24        $X_{1\ldots S,\mathrm{best}} \leftarrow X'_{1\ldots S}$

25        $d_{\mathrm{best}} \leftarrow \vec{d}(k)$

```
26       end if
27
```

28   `if` $\vec{d}(k) \equiv L$

29        $\vec{d}(k) \leftarrow d_{\mathrm{best}}$

30   `if` $k < |\vec{d}|$

31          $k \leftarrow k+1$

```
32          else
33             exit loop
34          end if
35       end if
```

36      $\vec{d}(k) \leftarrow \vec{d}(k) + 1$

```
37   end loop
```

is performed which gets the problem pipeline sets $X_{1...S}$ and returns the completed pipeline sets $X'_{1...S}$ for a given decision vector $\vec{d}$. The RPAG depth search is given in Algorithm 4.3 which contains only small extensions of the basic RPAG algorithm (Algorithm 4.1) to support the decision vector. Then, it is checked if the found solution is better than any solution so far and its solution as well as the decision is stored in d_{best} (lines 22-26). Now, the decision vector is adjusted in lines 28-37 by either scanning the solutions of one decision (line 36) or fixing one decision and moving to the next decision (line 29-34). The loop terminated when the last decision was evaluated (line 33).

The complete search is illustrated using an PMCM optimization with a simple example target set $T_{\text{odd}} = \{63, 173\}$ and $\text{cost}_A(x) = \text{cost}_R(x) = 1$. Figure 4.3 shows the decisions from top to bottom which are taken during the optimization for $L = 1$, i. e., the first and second best predecessors are evaluated in each selection. The order of the different depth search solutions as well as their decision vectors are indicated at the bottom from left to right. The algorithm starts from stage $S = \max(\text{AD}_{\text{min}}(63), \text{AD}_{\text{min}}(173)) = 3$ and searches for predecessors in stage 2. The working set is initialized to the odd target values ($W = \{63, 173\}$). From this working set, the potential predecessors 1, 7, 9, 21 and 63 from topologies (a) and (b) are found. Any of them can be used to compute element 63 from the working set, thus, the gain is one for all predecessors.

In the first depth search (leftmost path in Figure 4.3), the one with least magnitude is chosen for the predecessor set ($P = \{1\}$) and '63' is removed from the working set. In the next iteration, the possible predecessors '43', '45', '83', '87' (and more) are found from topology (c). Again, as the gain is one for all of them, '43' is chosen as it has the smallest magnitude and '173' is removed from the working set. Now, the working set is empty and the search for predecessors in stage one can be started by moving the elements from the predecessor set to the working set. The only possible single predecessor is element '1' from topology (a). Hence, it is selected and '43' remains in the working set. Unfortunately, '43' can not be computed from element '1' and a single additional predecessor with adder depth one. Hence, the predecessor pair search finds potential predecessor pairs, all of them with a gain of $\frac{1}{2}$. The pair with the lowest average magnitude is selected ($\{3, 5\}$) and the first (greedy) search is complete with a solution of seven nodes in the pipelined adder graph.

In the second depth search (second left path in Figure 4.3), the second best predecessor ('7') is selected in the first decision (according to

Algorithm 4.3: A single depth search within the RPAG (v2) algorithm

```
1  RPAG_DS (X₁...ₛ ,d⃗,S)
2    k := 0
3    for s = S downto 2
4      P := Xₛ₋₁
5      W := Xₛ \ 𝒜*(P)
6      while |W| ≠ 0 do
7        if d⃗(k) = ∅
8          d⃗(k) = 0
9        end if
10       p ← best_single_predecessor(P, W, s, d⃗(k))
11       if p ≠ 0
12         P ← P ∪ {p}
13       else
14         (p₁, p₂) ← best_predecessor_pair(W, s, d⃗(k))
15         P ← P ∪ {p₁, p₂}
16       end if
17       k ← k + 1
18       W ← W \ 𝒜*(P)
19     end while
20     Xₛ₋₁ ← P
```

$\vec{d} = (1, 0, 0, 0)^T$). The following decisions are again taken using a greedy approach, leading to a solution with six nodes in the pipelined adder graph, one less than in the first depth search. As $L = 1$, the algorithm stops to evaluate the first decision and fixes predecessor '7' in this stage for the following iterations.

The same procedure is repeated as shown in Figure 4.3, leading to several different solutions. The first best solution with $X_2 = \{7, 45\}$ and $X_1 = \{1, 3\}$ is finally selected. In this example, the second depth search already found the best solution. Of course, using more realistic cost functions in the low-level cost model will lead to a much higher probability to find better solutions in later depth search iterations. Note that the corresponding elements in X_s can be stored as soon as a decision is fixed to speed up the optimization. In the example above, after fixing predecessor '7' in the second depth search, it can be included in X_2. This reduced the problem size with every fixed decision.

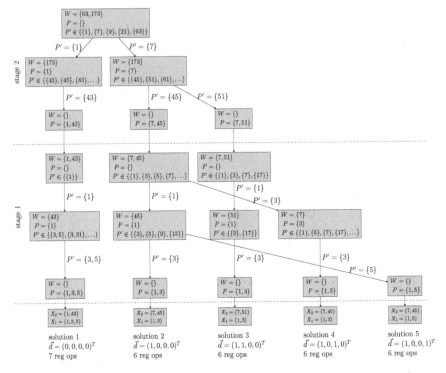

Figure 4.3: Decision tree of the overall RPAG algorithm for $T_{\text{odd}} = \{63, 173\}$ and $L = 1$

4.2.6 Computational Complexity

The computational complexity of the RPAG algorithm depending on the maximum word size B_T of the target set and the search limit L is derived in the following. This can be split into the overall algorithm and the evaluation of predecessor values.

In a single depth search of the overall algorithm, all elements in the pipeline sets X_s are evaluated in the worst case except for stage $s = 1$ (here, the predecessor '1' is already known). Thus, the number of elements to evaluate is equal to the sum of the set sizes $|X_s|$ for $s = 2 \ldots S$. Clearly, the last stage (S) is limited by the size of the odd target set $M = |T_{\text{odd}}|$, i. e., $|X_S| = M$. The previous stage can have at most twice the number of elements (then, no predecessors are shared), and so on. Hence, the total number of elements to evaluate in a single depth search ($L = 0$) is bounded by

$$E_0 = \sum_{s=2}^{S} |X_s| \leq \sum_{s=2}^{S} 2^{S-s} M = M(2^{S-1} - 1) \,. \tag{4.25}$$

The number of stages S is equal to the maximum of the minimal AD of the elements in the target set. This directly depends on the maximum number of non-zeros according to (2.18). A B bit MSD coded number can have up to $\lfloor B/2 \rfloor + 1$ non-zeros in the worst case (then every second bit is non-zero except for odd B where the two most significant bits are both non-zero in addition), leading to a maximum adder depth of

$$D_{\max} = \log_2(\lfloor B/2 \rfloor + 1) \,. \tag{4.26}$$

If we assume that B is a power-of-two, and the pipeline depth is chosen to be minimal, the number of stages is

$$S = \log_2(B/2 + 1) \tag{4.27}$$

Inserting this into (4.25) results in

$$E_0 \leq M(B/4 - \frac{1}{2}) \in \mathcal{O}(BM) \,, \tag{4.28}$$

where $\mathcal{O}$ denotes the Landau symbol (the "big-O" notation) [135].

For $L > 0$, there follow L depth searches with E_0 elements, then L searches with $E_0 - 1$ elements, etc., leading to

$$E_L = E_0 + L \sum_{e=0}^{E_0} e = E_0 + \frac{1}{2} L E_0(E_0 + 1) \in \mathcal{O}(LE_0^2) \subseteq \mathcal{O}(LB^2M^2) \tag{4.29}$$

elements in total. In the worst case, all topologies in Figure 4.2 have to be evaluated for each of these working set elements. The complexity for evaluating each topology is derived in the following.

The number of predecessor elements that can occur in topology (a) according to (4.7) is limited by the maximum size of the working set W. Assuming again that the predecessor set size can be at most twice the working set size, the worst case working set size is reached in stage 1, leading to:

$$|W| \leq 2^{S-1} M = BM/4 + M/2 \subseteq \mathcal{O}(BM) \tag{4.30}$$

Hence, topology (a) has a computational complexity of $\mathcal{O}(BM)$.

In topology (b), each working set element is divided by $2^k \pm 1 < 2^B$, according to (4.8), leading to $2B$ additional iterations compared to topology (a) and a complexity of $\mathcal{O}(MB^2)$.

In topology (c), all elements in the working set are combined with those in the predecessor set. As the predecessor set is limited by the same bound as the node set, the remaining complexity is $\mathcal{O}(B^2M^2)$.

In topology (d), all permutations of two elements of the working set are evaluated, leading to $\binom{|W|}{2}$ combinations. For each permutation, two loops up to a maximum shift $k_{max} \leq B + 2$ (see Appendix B) have to be evaluated for each of the cases that are described in Section 4.2.3. Hence, the complexity to compute topology (d) is

$$k_{max}^2 \binom{|W|}{2} \leq (B+2)^2 \binom{BM/4 + M/2}{2} \tag{4.31}$$

$$= (B+2)^2/2 \left((BM/4 + M/2)^2 - BM/4 - M/2 \right) \tag{4.32}$$

$$\in \mathcal{O}(B^4M^2) \tag{4.33}$$

The computationally most demanding part is the evaluation of permutations of predecessor pairs in topology (e). Here, all permutations of $\mathrm{nz}_{max}/2$ non-zeros out of nz_{max} non-zeros are computed in the worst case, leading to a complexity of

$$\binom{\mathrm{nz}_{max}}{\mathrm{nz}_{max}/2} = \frac{\mathrm{nz}_{max}!}{(\mathrm{nz}_{max}/2)!^2} \,. \tag{4.34}$$

It follows from the definition of the binomial coefficient [135] that this term is less than $2^{\mathrm{nz}_{max}}$ but greater than $2^{\mathrm{nz}_{max}/2}$. Although this is still an exponential growth it is much less than a complete search in which all the numbers between $1 \ldots 2^B$ have to be evaluated. Concrete values for the complexity of the MSD evaluation depending on the number of adder stages as well as a full search are listed in Table 4.1. It can be seen that up to four adder stages, the complexity is quite feasible (i. e., computable within few seconds) while it literally explodes for more stages. Four stages are enough to find solutions for coefficient word sizes up to $B = 31$ bits which is usually more than needed in most applications. Nevertheless, for very large problems the implementation can be forced to limit the number of MSD combinations to a pre-defined limit.

Table 4.1: Complexity of the MSD evaluation in topology (d) compared to a full search

Stages (S)	max. non-zeros $(\text{nz}_{max} = \lfloor B/2 \rfloor + 1)$	max. word size (B)	$\binom{\text{nz}_{max}}{\text{nz}_{max}/2}$	full search 2^B
1	2	3	$\binom{2}{1} = 2$	8
2	4	7	$\binom{4}{2} = 6$	128
3	8	15	$\binom{8}{4} = 70$	32768
4	16	31	$\binom{16}{8} = 12870$	$2 \cdot 10^9$
5	32	63	$\binom{32}{16} = 6 \cdot 10^8$	$9 \cdot 10^{18}$
6	64	127	$\binom{64}{32} = 2 \cdot 10^{18}$	$2 \cdot 10^{38}$

4.3 Lower Bounds of Pipelined MCM

Lower bounds for the objective of an optimization problem are useful to prove the optimality of a heuristic solution. A heuristic solution with an objective identical to a known lower bound is optimal whereas the optimality is still unknown if the lower bound is not reached. Lower bounds for the number of adders of non-pipelined SCM, MCM and CMM problems were proposed by Gustafsson [74]. The initial lower bound of adders for the MCM problem is given as

$$A_{\text{LB,MCM}} = N_{\text{uq}} + \min_{t \in T}(\text{AD}_{\min}(t)) - 1 \qquad (4.35)$$

where N_{uq} is the number of unique, odd target coefficients (except fundamental '1' as it requires no adder to realize). This bound can be further increased if there are gaps in the adder depths of the target set T, e. g., if there are only coefficients with adder depth one and three, there is at least one additional adder necessary to realize a coefficient with adder depth two (formal equations can be found in [74]). This bound is still valid for the PMCM problem. However, the primary goal in PMCM is the number of registered operations (including registers). For PMCM we will have exactly N_{uq} coefficients in the last stage (now including fundamental '1' in N_{uq} as it requires at least a register), whereas in each previous stage there has to be at least one element, leading to a simple lower bound of

$$O_{\text{LB,PMCM}} = N_{\text{uq}} + S - 1 \ . \qquad (4.36)$$

This bound is already larger than or equal to (4.35). However, this bound can be further improved by the fact that an element with an adder depth of S (at least one exists by definition) can only be computed by a single predecessor in stage $S - 1$ by topology (b) in Figure 4.2 (see page 59). So if there is any element in the target set for which this condition fails (e. g., a prime number) the lower bound can be increased by one. Furthermore, for $|T_{\text{odd}}| > 1$, topology (d) can be used to check if more than two elements are necessary in stage $S-1$. Only if all remaining target coefficients that can not be realized by a single predecessor using topologies (a)-(c) can be computed from topology (d), two elements are sufficient, otherwise, the lower bound can be again increased by one. However, there is no closed form for this improved lower bound, so it has to be evaluated numerically.

4.4 Implementation

RPAG was implemented as a flexible command-line tool in C++ and is freely available as open-source [16]. It heavily uses the C++ standard template library (STL) and itself is template based for supporting different problem extensions (see chapters 6, 8 and 10). The current version (1.5) consists of 6000 lines of code (including all extensions). It can be compiled by different C++ compilers (tested with gcc [136] and clang [137] using Linux, MacOS and Windows) without the use of any additional library (besides the standard C++ libraries).

The only mandatory argument is a list of constants, e. g., calling

```
rpag 63 173
```

optimizes the PMCM problem for the target set $\{63, 173\}$ resulting in

```
best result: pag_cost=6, pipeline_set_best={[1 3],[7 45],[63 173]}
```

meaning that the solution pipeline sets are equal to $X_1 = \{1, 3\}$, $X_2 = \{7, 45\}$ and $X_0 = \{63, 173\}$. Providing the command line flag --show_adder_graph also provides the information for the construction of the adder graph:

```
pipelined_adder_graph={{1,1,1,0,0,0,0,0},{3,1,1,0,0,1,0,1},
{7,2,1,1,3,-1,1,0},{45,2,3,1,4,-3,1,0},{63,3,7,2,0,7,2,3},
{173,3,45,2,2,-7,2,0}}
```

Each node in curly brackets is in the format $\{w, s'_w, u, s'_u, l'_u, v, s'_v, l'_v\}$, where s'_w, s'_u and s'_v are the pipeline stages of w, u and v, respectively (for PMCM they are always related by $s_w = s_u + 1 = s_v + 1$). The other parameters are already known from the $\mathcal{A}$-operation as defined in Section 2.4.1. Registers are encoded with $w = u$ and $v = l'_u = l'_v = 0$. Node {173,3,45,2,2,-7,2,0}, for example, is computed in stage '3' as $173 = 45 \cdot 2^2 - 7$.

The syntax is compatible to a Matlab cell array and Matlab scripts are provided with RPAG that automatically generate pdf-plots using the graphviz graph visualization software [138]. These graphs were used throughout the thesis and are looking like those already shown in Figure 4.1.

There are many flags that influence the optimization. The cost functions as well as the problem type are set by --cost_model, which can be, e.g., hl_fpga, ll_fpga or min_ad for the high/low level FPGA cost models and the MCM$_{\text{MAD}}$ problem, respectively. For the low-level cost models, the input word size has to be provided by --input_wordsize. The search limit can be set by --search_limit (default: 10). A deeper pipeline (compared to the minimal possible S) can be set by specifying --no_of_extra_stages. To understand the process of the optimization, different verbose levels are implemented which can be set by --verbose. A comprehensive list of command line arguments are obtained by calling rpag --help.

4.5 Experimental Results

Different experiments were performed to analyze the algorithm and its parameters and to compare the algorithm to state-of-the-art methods which are presented in the following subsections.

4.5.1 Registered Operations and CPU Runtime

To get an impression about the behavior of the number of registered operations and runtime of the algorithm for different numbers of coefficients and coefficient word sizes, a benchmark with a statistical relevant number of problem instances was used. The number of coefficients N was varied between 1 and 100 where for each N, 100 random coefficients sets with a word size of $B_c \in \{8, 16, 24, 32\}$ (uniformly distributed between 1 and $2^{B_c} - 1$) were optimized. The high-level cost model was used $(\text{cost}_A(x) = \text{cost}_R(x) = 1)$ and the search limit L was set to 0 in this experiment. The average CPU runtime was measured on a single CPU of an Intel Nehalem 2.93 GHz computer.

The number of registered operations and runtime versus the number of coefficients is plotted in Figure 4.4(a) and Figure 4.4(b), respectively. The same data is shown for varying coefficient word size in Figure 4.4(c) and Figure 4.4(d). It can be observed that the registered operations roughly follow a linear trend while the runtime grows exponentially. However, the runtime is less than 35 seconds on average even for the largest problem. Note that it can happen that less registered operations than coefficients are required for low coefficient word sizes. This is due to the fact that the number of unique coefficients N_{uq} may be much less compared to N in this case (e. g., with $B_c = 8$ bit and $N = 100$, there are only 100 random values out of 255, so many duplicates may occur).

4.5.2 Influence of the Search Width Limit to the Optimization Quality

To obtain better optimization results by spending more runtime, the search width limit L is a crucial parameter. Therefore, experiment of the previous section was repeated with $B_c = 16$ bit and varying $L \in \{0, 1, 2, 10, 50\}$. The average number of registered operations is shown in Figure 4.5(a) for different L. The absolute as well as the percentage improvement of registered operations compared to $L = 0$ is shown in Figure 4.5(b) and Figure 4.5(c), respectively. Note that a pure greedy search was proposed in the original RPAG publication [8], corresponding to $L = 0$. The average CPU time is shown in Figure 4.5(d).

As expected, the higher L, the lower the average number of registered operations. More registered operations can be saved with increasing N. The percentage improvement is highest for low N reaching an improvement of up to 32% for $N = 1$ and $L = 50$. The required computation time is only a few seconds for low L. The average runtime for $L = 50$ reached 12 minutes for $N = 100$. Setting the search width limit to $L = 10$ seems to be a good compromise between runtime and optimization quality. Even in the worst case, less than five minutes were required for 16 bit coefficients which is reasonable compared to the synthesis time.

4.5.3 Comparison of Cost Models

In this experiment, the high-level and low-level cost models are compared. The same data and settings as in the last section were used with $L = 0$

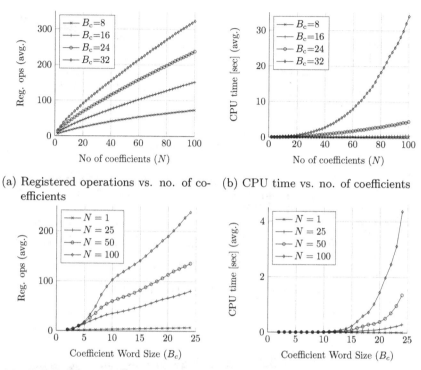

(a) Registered operations vs. no. of coefficients

(b) CPU time vs. no. of coefficients

(c) Registered operations vs. coefficient word size

(d) CPU time vs. coefficient word size

Figure 4.4: Optimization results for different N and B_c with $L = 0$

and $L = 10$. Figure 4.6 shows the resulting low-level cost improvement of the optimization result for the low level cost modes (using (2.26) and (2.27) with $K_{FF} = 1$ and $B_i = 16$ bit) compared to the high level cost as RPAG optimization goal. It can be observed that the improvement can be large for a few coefficients but not more than 1% can be gained on average using the detailed cost model. The reason for this relatively low improvement is the preference of predecessors with low magnitude which reduces the low-level costs even in the high-level cost model.

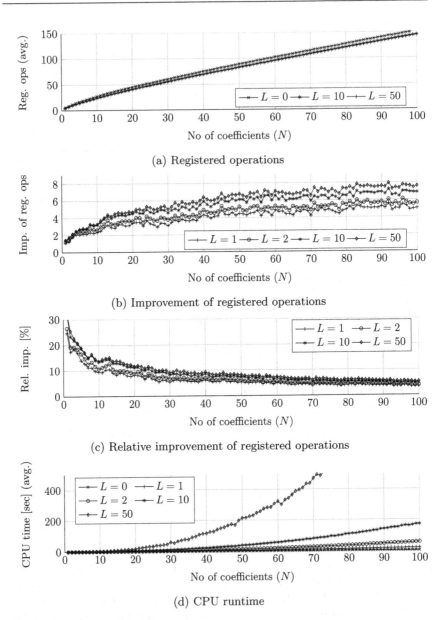

(a) Registered operations

(b) Improvement of registered operations

(c) Relative improvement of registered operations

(d) CPU runtime

Figure 4.5: Optimization results for different search width limits L and $B_c = 16$ bit

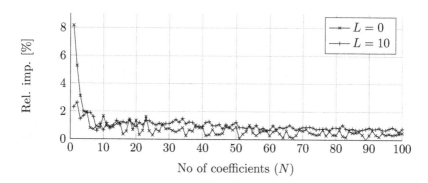

Figure 4.6: Resulting low-level improvement when optimized with the low-level cost model compared to the high-level cost model

4.5.4 Comparison Between Gain Functions

Another improvement compared to the original RPAG as proposed in [8] is the extension of the gain function. In the original gain function

$$\text{gain}(P') = \sum_{w \in W_R(P')} 1/\text{cost}_R(w) + \sum_{w \in W_A(P')} 1/\text{cost}_A(w) \, , \qquad (4.37)$$

only the cost of working set elements were considered instead of considering the relation of predecessor cost to the cost of realized elements from the working set as done in (4.24).

The effect of the new gain function is validated in an experiment using the same benchmark and settings as in Section 4.5.2 but with the low-level cost function and $L = 0$ and $L = 10$. The average of the low-level cost is shown in Figure 4.7(a). The average absolute and relative improvements of the new gain function compared to the old gain function are shown in Figure 4.7(b) and Figure 4.7(c), respectively. It can be observed that an almost constant relative improvement between 2-4% is achieved with the new gain function, on average. The only exception is for $L = 0$ with $N = 1$ (the SCM case) and $N = 2$, where the improvement is about 8% and 5%, respectively. Of course, there are cases where the new gain function may lead to a worse result but on average, it clearly leads to a improvement in the optimization quality.

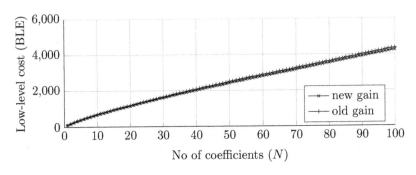

(a) Low-level cost

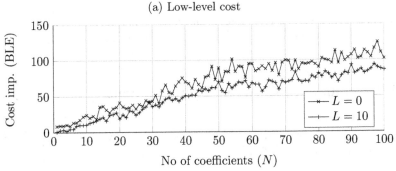

(b) Improvement of low-level cost using the new gain function

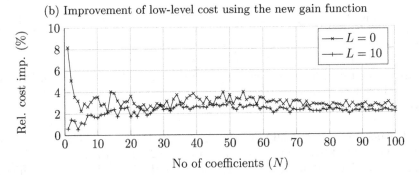

(c) Relative improvement of low-level cost using the new gain function

Figure 4.7: Comparison of different gain functions for the low-level cost model

4.5.5 Influence of the Pipeline Stages to the Optimization Quality

An important parameter of the PMCM problem is the pipeline depth S. As each additional pipeline stage may introduce many additional adders, it was assumed that it is unlikely that an additional stage leads to a better result. Hence, the pipeline depth was defined to be the minimum possible, which is only limited by the maximum adder depth D_{max}. To evaluate the influence of the pipeline depth, the experiment of Section 4.5.2 was repeated with $L = 0$ and $L = 10$ and one additional pipeline stage (i. e., $S := D_{max} + 1$). As expected, typically more adders are needed on average leading to a negative improvement compared to the minimum depth case as shown by the curves marked with circle and '+' in Figure 4.8(a). However, there are cases in which an additional stage improves the result. Hence, the improvement was computed if the best out of both results for $S \in \{D_{max}, D_{max} + 1\}$ is taken compared to the $S = D_{max}$ case. These results are indicated as "best of both" in Figure 4.8(a) (curves marked with '×' and diamond) Obviously, the "best of both" is better or equal per definition. It leads to an improvement of up to 0.75 registered operation on average and a relative improvement in the order of $0.25\ldots3.75\%$ (see Figure 4.8(b)). Of course, the improvement may be much higher than the average. The highest improvement was five registered operations in this benchmark (for $L = 10$). A histogram showing the number of cases where an instance with one extra stage was better than the minimum pipeline depth is shown in Figure 4.9. It can be observed that the chance is always less than 30% to get a better result but may be high enough to be worth to evaluate.

To get an idea about the reason of such an improvement, Figure 4.10 shows the pipelined adder graphs for a 12 coefficient instance where the improvement was maximal. While 32 registered operations are required for minimum pipeline depth only 28 are needed with one extra stage. It can be seen that 13 operations are required in the second last stage for minimum pipeline depth while only eight are required in the other case. This reduction can be explained by the fact that there are much more possible predecessors with an adder depth of three compared to those with an adder depth of two (more details can be found in Section 5.5). In this example there are only three predecessors (out of eight) with an adder depth of three which are used to reduce the predecessor count from 13 to 8 in this stage. This comes at the expense of a slight increase of previous operations (eight operations distributed over two stages instead of seven in a single stage) but still leading

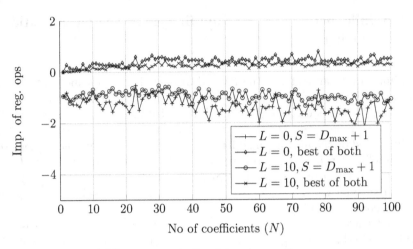

(a) Improvement of registered operations

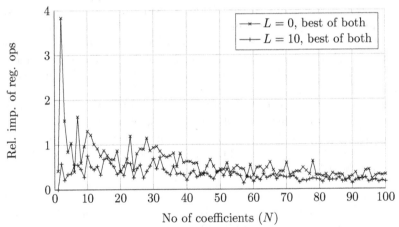

(b) Relative improvement of registered operations

Figure 4.8: Average improvement of registered operations when using one extra stage ($S = D_{max} + 1$) compared to $S = D_{max}$ and the best of both result for $L = 0$ and $L = 10$

to a reduction in total.

Another example is the pipelined SCM shown in Figure 4.11. Here, the coefficient can be factored into $26565 = 805 \cdot 33 = 161 \cdot 5 \cdot 33$, where 5 and 33

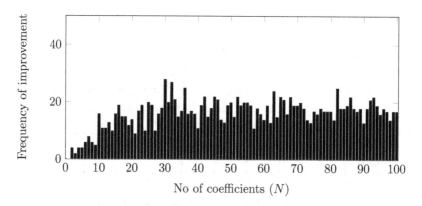

Figure 4.9: Histogram of number of instances (out of 100) which are better when using one extra stage $(S = D_{\max} + 1)$ compared to $S = D_{\max}$

are cost-1 coefficients (numbers that can be realized by a single adder from '1'). All resulting predecessors according to topology (b) (see Figure 4.2 on page 59) have an adder depth of three. Hence, with minimal pipeline depth, the only way is using two elements in the second last stages whereas one element is sufficient with one extra stage. The shown adder graph with increased pipeline depth is an example where an additional stage reduces the operations. Note that there were only 2 out of 100 SCM instances were an extra stage leads to an improvement.

4.5.6 Comparison with the Optimal Pipelining of Precomputed Adder Graphs

To compare the quality of the RPAG algorithm with previous MCM methods and the optimal pipelining as presented in Chapter 3, the same benchmark as in Section 3.3 is used, together with the H_{cub} algorithm targeting a bounded adder depth ($H_{\text{cub,BAD}}$). In this experiment, the high-level FPGA model was used. The results are listed in Table 4.2. All computations were done on a notebook with 2.4 GHz Intel Core i5 CPU. Results from the original RPAG [8], denoted as RPAG (v1), are also given to show the influence of the proposed improvements, denoted as RPAG (v2). Note that the RPAG (v1) results in [8] were obtained by using $R = 50$ runs of the algorithm where predecessors with identical gain are randomly selected. This non-determinism has the disadvantage that results are hard to reproduce. To have a fair com-

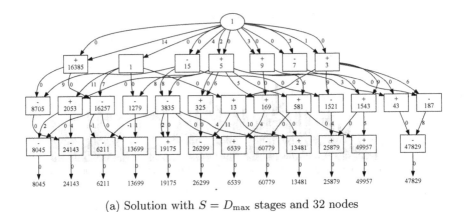

(a) Solution with $S = D_{\max}$ stages and 32 nodes

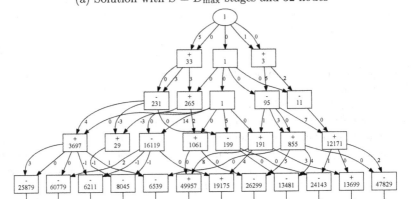

(b) Solution with $S = D_{\max} + 1$ stages and 28 nodes

Figure 4.10: Example PAGs where an increased pipeline depth is beneficial

parison, 50 runs per filter with random selection (flag 'nofs=false') were also used for $H_{\text{cub,BAD}}$ and RPAG (v2) was used with $L = 50$.

It can be observed that the number of registered operations can be further reduced for all benchmark instances compared to the $H_{\text{cub,BAD}}$ with the optimal pipelining as proposed in Section 3. On average, a reduction of 11.5% of registered operations is obtained with the proposed RPAG (v2) algorithm. Compared to the RPAG (v1) algorithm, the average improvement is 1.65%. All solutions could be found within a few seconds. This reduction is on top

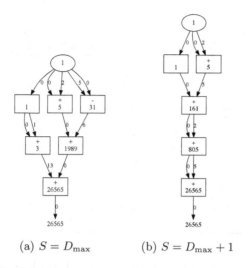

(a) $S = D_{\max}$ (b) $S = D_{\max} + 1$

Figure 4.11: Pipelined adder graph SCM example where an increased pipeline depth is beneficial

Table 4.2: Comparison between RPAG and the optimal pipelining of precomputed adder graphs obtained by $H_{cub,BAD}$ based on benchmark filters from [5]. Abbreviations: N_{uq}: no. of unique coefficients, reg. add: registered adder/subtractor, pure regs: registers, reg. ops: registered operations

		$H_{cub,BAD}$ opt.pip.			RPAG (v1) $R = 50$ [8]				RPAG (v2)			
N	N_{uq}	reg. add	pure reg.	reg. ops	reg. add	pure reg.	reg. ops	comp. time [s]	reg. add	pure reg.	reg. ops	comp. time [s]
6	3	6	4	10	8	1	**9**	0.20	6	3	**9**	0.02
10	5	9	5	14	10	3	13	0.37	9	3	**12**	0.14
13	7	12	4	16	14	2	16	0.42	15	0	**15**	0.24
20	10	15	6	21	14	4	**18**	0.51	15	3	**18**	0.50
28	14	20	6	26	20	3	**23**	0.62	20	3	**23**	0.99
41	21	25	13	38	31	1	**32**	1.30	31	1	**32**	2.64
61	31	35	12	47	39	3	42	1.93	38	3	**41**	4.2
119	54	53	23	76	62	7	69	4.92	64	3	**67**	9.90
151	71	73	18	91	79	4	**83**	4.60	79	4	**83**	8.57
avg.:	24.0	27.6	10.1	37.7	30.8	3.1	33.9	1.65	30.8	2.6	33.3	3.02
imp.:								10.0%			**11.5%**	

Table 4.3: Comparison of the proposed method for the low-level ASIC model with HCUB-DC+ILP [7] based on benchmark filters from [7].

Filter	HCUB-DC+ILP [7]				RPAG (v1) $R = 50$ [8]			
ID	reg. add	pure reg.	reg. ops	icost	reg. add	pure reg.	reg. ops	icost
A30	24	8	32	64,698	24	5	29	**52,958**
A80	54	7	61	143,666	57	5	62	**131,150**
A60	39	15	54	112,922	43	5	48	**97,350**
A40	30	7	37	81,186	30	5	35	**68,724**
B80	45	12	57	117,614	46	7	53	**111,392**
A300	82	61	143	255,076	87	10	97	**243,642**
avg.:	45.67	18.33	64	129,194	47.83	6.17	54	117,536
imp.:							15.63%	9.02%

of the 54.1% (slice) improvement compared to the add/shift method [6] as obtained in Section 3.3.2.

4.5.7 Comparison Between Aksoy's Method and RPAG

In this section, the RPAG algorithm is compared to the pipelined MCM method of Aksoy et al. [7] which is called HCUB-DC+ILP (see Section 2.6.4 for a brief description of this method). They used a low-level cost model for the implementation cost (icost) of ASICs [121]. To compare with them, the same benchmark filters were used and the same cost model was applied to the resulting adder graphs. All coefficient values are also available online in the FIR benchmark suite [127] labeled as AKSOY11_ECCTD_ID. The results are listed in Table 4.3. Less resources could be obtained in all cases by using RPAG (the best are marked bold) compared to Aksoy's method, leading to an average reduction of 9%. Note that the original RPAG algorithm [8] was used in this experiment without further extensions as the icost model as provided from the author of [7] was implemented in Matlab like the original RPAG implementation.

4.6 Conclusion

A heuristic for solving the PMCM problem was proposed in this chapter. It is a fast and flexible algorithm which is fairly simple and produces better results than the state-of-the-art MCM algorithm H_{cub} [41] combined with the optimal pipelining of Chapter 3 or state-of-the-art PMCM heuristics proposed previously [6, 7]. Its flexibility will be further used in this work for related problems, where the RPAG algorithm will serve as a base. In particular RPAG is extended for the pipelined CMM (PCMM) and the CMM with minimal AD problems (Chapter 6), FPGA-specific problems like the inclusion of embedded multipliers in the PMCM problem (Chapter 8) and the PMCM problem with ternary adders (Chapter 10). Before that, we will consider how to optimally solve the PMCM problem using ILP methods in the following. Optimal approaches will always be limited to small problem instances but gives valuable insights when compared to heuristic solutions. So it will be used to further quantify the optimization quality of the RPAG heuristic for small problems but will be also used to solve practical relevant problems from image and video processing.

5 Optimally Solving MCM Related Problems using Integer Linear Programming

In this chapter, optimal methods are considered to solve the PMCM problem as well as the MCM problem with adder depth constraints (the MCM_{MAD} and MCM_{BAD} as defined in Section 2.4). For that we formulate the problems using integer linear programming (ILP). Of course, optimal results can only be achieved for relatively small problems, so there will be still a need for good heuristics like the RPAG algorithm presented in the last chapter. However, practical problems exist that can be solved optimal and from the obtained solutions it is sometimes possible to reveal the weak spots in the heuristic. The main ideas in this chapter were originally published in [139] but are further extended in this chapter.

5.1 Related Work

A quite flexible ILP formulation for the MCM problem was proposed by Gustafsson [97]. The main idea is to represent the MCM search space as a directed hypergraph in which the solution is a Steiner hypertree [105] connecting the input node with all nodes that correspond to target values. A hypergraph is a generalized graph in which each edge can connect more than two nodes. Like for an adder graph, each node in the directed adder hypergraph corresponds to an odd fundamental, but in contrast each hyperedge has two tail nodes and one head node. If a fundamental w can be obtained from fundamentals u and v using one $\mathcal{A}$-operation ($w = \mathcal{A}_q(u, v)$) the corresponding nodes are connected by a hyperedge $(u, v) \rightarrow w$. A simple example is given in Figure 5.1(a), where a small part of the search space containing the nodes of the target set $T_{\text{odd}} = \{11, 13, 35\}$, the input node '1' and the three NOFs '3', '7' and '9' with their corresponding hyperedges are illustrated. As each node in the hypergraph corresponds to an adder, the

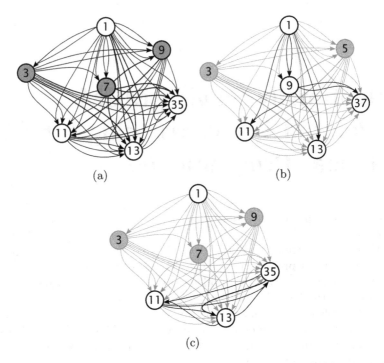

Figure 5.1: Example hypergraphs for target set $T = \{44, 70, 104\}$ (a) subset of the search space including the Steiner nodes 3, 7 and 9, (b) one possible solution hypertree by selecting 7 as Steiner node (c) an invalid solution containing loops

MCM problem can now be reformulated in finding a minimum Steiner tree in this hypertree which connects input node '1' and the odd target fundamentals in T_{odd} (the white nodes in Figure 5.1). All additional nodes that have to be included to build the Steiner tree are called Steiner nodes and directly correspond to the NOFs. One possible solution would be to select the Steiner node '7', resulting in the hypertree as highlighted in Figure 5.1(b). The complete hypergraph has to span the complete search space containing nearly all possible fundamentals from 0 to c_{max} (there may be less due to the elimination of nodes which are not connected to any target coefficient).

To solve the Steiner hypertree problem, an ILP formulation was proposed that is based on the Steiner tree formulation of Voss [140] and is given in ILP Formulation 3. It was kindly provided by Oscar Gustafsson as it was not printed in the original publication [97]. The set of vertex nodes is denoted with $V \subseteq \mathbb{N}$ and the set of hyperedges E contains all triplets (u, v, w) for which a $w = \mathcal{A}_q(u, v)$ exists.

For each node, the binary variable x_w is defined to be '1' if node w is part of the Steiner hypertree solution. Likewise, the binary variable $x_{(u,v,w)}$ defines the use of hyperedge $(u, v) \to w$ in the Steiner hypertree. Now, constraint C1 defines that all target nodes and the input node '1' have to be included in the Steiner hypertree. Constraint C2 ensures that there is exactly one edge going to each node in the Steiner hypertree. For each edge in the Steiner hypertree, the corresponding input nodes have to be available which is given by constraint C3.

ILP Formulation 3 (MCM Problem [97]).

$$\min \sum_{w \in V \setminus \{1\}} x_w$$

subject to

C1:
$$x_w = 1 \text{ for all } w \in T_{\text{odd}} \cup \{1\}$$

C2:
$$x_w - \sum_{(u,v,w') \in E \,|\, w'=w} x_{(u,v,w)} = 0 \text{ for all } w \in V$$

C3:
$$x_u + x_v \geq 2x_{(u,v,w)} \text{ for all } (u, v, w) \in E$$

C4a:
$$M\, x_{(u,v,w)} + d_u - d_w \leq M - 1 \text{ for all } (u, v, w) \in E$$

C4b:
$$M\, x_{(u,v,w)} + d_v - d_w \leq M - 1 \text{ for all } (u, v, w) \in E$$

C5:
$$d_1 = 0$$

C6:
$$d_w \leq 1 \text{ for all } w \in \{2^k \pm 1\} \cap V$$
$$\text{with } k \in \mathbb{Z}_0$$

C7:
$$d_w \geq 1 \text{ for all } w \in T_{\text{odd}} \setminus \{1\}$$
$$x_w \in \{0, 1\} \text{ for all } w \in V$$
$$x_{(u,v,w)} \in \{0, 1\} \text{ for all } (u, v, w) \in E$$
$$d_w \in \mathbb{N}_0 \text{ for all } w \in V$$

So far, the constraints guarantee that the solution will contain all targets and each target can be computed from other targets, the input or additional Steiner nodes. However, it may happen that cycles occur as illustrated in

Figure 5.1(c). It directly follows from Lemma 2 (see page 16) that if there is an edge from $(u, v) \rightarrow w$ then there is also an edge from $(w, v) \rightarrow u$ and $(u, w) \rightarrow v$. To avoid cycles, constraints C4a and C4b are used which are hypergraph extensions of the Miller-Tucker-Zemlin (MTZ) constraints [141], originally introduced for the sub-tour elimination in the traveling salesman problem. They ensure that the adder depth of node w, represented by integer variable d_w, is at least one more than the adder depth of nodes u and v (d_u and d_v) if the corresponding edge is used ($x_{(u,v,w)} = 1$), or is otherwise unconstrained by setting M to a sufficiently large number:

$$\left. \begin{array}{l} d_w \geq d_u + 1 \\ d_w \geq d_v + 1 \end{array} \right\} \text{ when } x_{(u,v,w)} = 1 \qquad (5.1)$$

$$\left. \begin{array}{l} d_w \geq d_u + 1 - M \\ d_w \geq d_v + 1 - M \end{array} \right\} \text{ when } x_{(u,v,w)} = 0 \qquad (5.2)$$

Constraints C6 and C7 are lifting constraints that are not necessary to get the correct solution but potentially reduce the optimization time. The common way to solve a (B)ILP problem is to iteratively solve and fix variables in the relaxed linear programming (LP) model (with real valued variables) in a branch-and-bound based fashion. Then, tighter bounds usually lead to improved run times during the (B)ILP optimization. The constraints are obtained by the observation that nodes directly computed from the input node '1' can not have a higher depth than one (constraint C6) and target coefficients must have a non-zero positive depth (constraint C7).

The formulation can be easily adapted to the MCM_{MAD} or the MCM_{BAD} problem by constraining the depth of the related coefficients, i. e., $d_t = \text{AD}(t)$ for the MCM_{MAD} problem and $d_t = D_{\max}$ for the MCM_{BAD} problem for all $t \in T_{\text{odd}}$.

Another optimization problem that is related to the PMCM problem is the set cover with pairs (SCP) problem, which was introduced by Hassin and Segev [142] and has applications in biological and biochemical problems [143]. In the SCP problem, a set of pairs $(i, j) \in A$ is given where each pair covers a subset V of a ground set U (i. e., $V \subseteq U$). Each pair is associated with a non-negative cost and the goal is to select a collection of pairs which covers all elements in U with minimum cost. An ILP formulation of the problem was presented by Gonçalves et al. [143]. The SCP problem can be used to solve the optimization of a single stage as performed in the RPAG algorithm (see Chapter 4). Then, each pair (i, j) corresponds to all possible pairs of predecessors in one stage $(i, j \in P)$ while the covered subset contains all elements that can be computed by one add/subtract operation or a register leading to $V = \mathcal{A}(i,j) \cup \{i,j\}$. The ground set U to be covered is identical to the working set W. However, it is an NP-hard problem [143] which is limited to the optimization of a single stage.

5.2 ILP Formulations of the PMCM Problem

In the following, ILP formulations are derived for the PMCM problem. They are based on ideas used in the adder graph pipelining in ILP Formulation 2 (see page 44) as well as the optimal MCM ILP Formulation 3. Three alternative ILP formulations are presented with different advantages/disadvantages. Their main difference compared to ILP Formulation 3 is the separation of fundamentals in different pipeline stages. This naturally occurs in the PMCM problem, as the same numeric value can be realized multiple times in different pipeline stages. This prevents any cycle and avoids the MTZ constraints as well as the depth variables. For that, some additional sets have to be defined as follows.

All possible odd fundamentals in pipeline stage s, denoted as $\mathcal{A}^s$, are obtained by recursively computing the $\mathcal{A}_*$ sets (see Definition 2 on page 17):

$$\mathcal{A}^0 = \{1\} \tag{5.3}$$
$$\mathcal{A}^s = \mathcal{A}_*(\mathcal{A}^{s-1}) \tag{5.4}$$

Not all of the elements of $\mathcal{A}^s$ may be needed to compute the target coefficients. Therefore, the set $\mathcal{S}^s \subseteq \mathcal{A}^s$ is defined which contains all *single* elements that may be used to compute the target coefficients T_{odd} in the last stage. In addition, $\mathcal{T}^s$ denotes the set of (u, v, w) *triplets* for which $w \in \mathcal{S}^s$ can be computed using u and v of the previous stage (i.e., $u, v \in \mathcal{S}^{s-1}$). Both sets $\mathcal{S}^s$ and $\mathcal{T}^s$ can be computed recursively, starting from the last stage S, where $\mathcal{S}^S$ is identical to the odd target coefficients excluding zero:

$$\mathcal{S}^S := \{\text{odd}(t) \mid t \in T \setminus \{0\}\} \tag{5.5}$$
$$\mathcal{T}^{s-1} := \{(u,v,w) \mid w = \mathcal{A}_q(u,v), \ u, v \in \mathcal{A}^{s-1}, \ u \leq v, \ w \in \mathcal{S}^s\} \tag{5.6}$$
$$\mathcal{S}^{s-1} := \{u, v \mid (u,v,w) \in \mathcal{T}^{s-1}\} \tag{5.7}$$

The sizes of these sets are analyzed in Section 5.5 as these directly correlates with the problem size.

5.2.1 PMCM ILP Formulation 1

The ILP formulation uses the same binary variables a_w^s and r_w^s as used in ILP Formulation 2 (see page 44) which are defined to be true if w is realized in stage s using an adder or a register, respectively. In addition, binary variables are introduced which determine if a pair of nodes (u, v) is available in stage s:

$$x_{(u,v)}^s = \begin{cases} 1 & \text{if both nodes } u \text{ and } v \text{ are available in stage } s \\ 0 & \text{otherwise} \end{cases} \tag{5.8}$$

Now, the ILP formulation to solve the PMCM problem can be formulated as given in PMCM ILP Formulation 1.

Cost functions $\text{cost}_A(w)$ and $\text{cost}_R(w)$ define the cost to realize w using an adder or a register, respectively (see Section 2.5.4). Constraints C1-C3 are equivalent to C2-C4 in ILP Formulation 2: Constraint C1 ensure that all target coefficients are present in the last stage while C2 and C3 require that a value w can only be stored in a register if it was computed before. Constraint C4 specifies that if w is computed by $w = \mathcal{A}(u, v)$ in stage s, the pair (u, v) has to be available in the previous stage. If a pair (u, v) has to be available in stage s then u and v have to be realized in that stage either as register or adder which is constrained by C5. Note that instead of using constraint C2 it will be more practical to remove all related variables r_w^s which are zero from the cost function and related constraints.

In the formulation, the relaxed condition $x_{(u,v)}^s \in \mathbb{R}_0$ is used instead of requiring $x_{(u,v)}^s \in \{0, 1\}$ which results in an mixed integer linear programming (MILP) formulation. Using continuous variables $x_{(u,v)}^s$ instead of binary ones typically reduces the runtime of the optimization substantially. This is possible since we can construct a binary solution that is feasible and minimal from the relaxed solution which is proven in the following.

PMCM ILP Formulation 1.

$$\min \sum_{s=1}^{S} \sum_{w \in \mathcal{S}^s} \left(\text{cost}_\text{A}(w) \, a_w^s + \text{cost}_\text{R}(w) \, r_w^s \right)$$

subject to

C1: $\qquad r_w^S + a_w^S = 1$ for all $w \in T_\text{odd}$

C2: $\qquad r_w^s = 0$ for all $w \in \mathcal{S}^s \setminus \bigcup\limits_{s'=0}^{s-1} \mathcal{S}^{s'}$

$\qquad\qquad$ with $s = 1 \ldots S - 1$

C3: $\qquad r_w^s - a_w^{s-1} - r_w^{s-1} \leq 0$ for all $w \in \mathcal{S}^s,\ s = 2 \ldots S$

C4: $a_w^s - \sum\limits_{(u,v,w') \in \mathcal{T}^s \,|\, w'=w} x_{(u,v)}^{s-1} \leq 0$ for all $w \in \mathcal{S}^s, s = 2 \ldots S$

C5: $\qquad \left.\begin{matrix} x_{(u,v)}^s - r_u^s - a_u^s \leq 0 \\ x_{(u,v)}^s - r_v^s - a_v^s \leq 0 \end{matrix}\right\}$ for all $(u,v,w) \in \mathcal{T}^s$
$\qquad\qquad$ with $s = 1 \ldots S - 1$

$\qquad\qquad a_w^s, r_w^s \in \{0,1\}$ for all $w \in \mathcal{S}^s,\ s = 1 \ldots S$

$\qquad\qquad x_{(u,v)}^s \in \mathbb{R}_0$ for all $(u,v,w) \in \mathcal{T}^s$

$\qquad\qquad$ with $s = 1 \ldots S - 1$

Proof. Assume an optimal solution of PMCM ILP Formulation 1 is given. Let w' be included in the solution which is realized as adder in stage s (i. e., $a_{w'}^s = 1$). Then, constraint C4 would allow several $x_{(u,v)}^s$ variables for which $(u,v,w') \in \mathcal{T}^s$ to lie in the range $0 \ldots 1$. Any $x_{(u,v)}^s$ variable greater than zero would lead to $r_u^{s-1} = 1$ or $a_u^{s-1} = 1$ and $r_v^{s-1} = 1$ or $a_v^{s-1} = 1$ due to constraint C5. There will be at least one pair (u',v') with minimal cost contribution in the solution:

$$(u',v') = \arg \min_{(u,v) \in \mathcal{P}_w} \left(\text{cost}_\text{A}(u)\, a_u^s + \text{cost}_\text{R}(u)\, r_u^s + \text{cost}_\text{A}(v)\, a_v^s + \text{cost}_\text{R}(v)\, r_v^s \right) \tag{5.9}$$

The cost contribution of all the corresponding (u,v) pairs to realize w' is given as:

$$\sum_{(u,v,w)\in\mathcal{T}^s \mid w=w'} \left(\mathrm{cost_A}(u)\, a_u^s + \mathrm{cost_R}(u)\, r_u^s + \mathrm{cost_A}(v)\, a_v^s + \mathrm{cost_R}(v)\, r_v^s\right)$$

$$\geq \mathrm{cost_A}(u')\, a_{u'}^s + \mathrm{cost_R}(u')\, r_{u'}^s + \mathrm{cost_A}(v')\, a_{v'}^s + \mathrm{cost_R}(v')\, r_{v'}^s \quad (5.10)$$

As the objective is to minimize the cost function, only one variable $x_{(u',v')}^s$ that leads to the minimal cost contribution is selected which forces all other variables $x_{(u,v)}^s$ that are not part in the optimal solution to be zero. The derivations above can be done for all $a_w^s = 1$ which shows that relaxing the variables $(x_{(u,v)}^s \in \mathbb{R}_0)$ leads to the same result as for $x_{(u,v)}^s \in \{0,1\}$. $\quad \Box$

5.2.2 PMCM ILP Formulation 2

One disadvantage in the first PMCM ILP formulation is the fact that a precise low-level cost model including right shifts is not possible. For that, the complete (u,v,w) vector has to be considered in the cost function. Therefore, the variable $x_{(u,v)}^s$ of the last formulation is extended similar to the hyperedge variables in ILP Formulation 3:

$$x_{(u,v,w)}^s = \begin{cases} 1 & \text{if } w \text{ in stage } s \text{ can be obtained from } u \text{ and } v \text{ in stage } s-1 \\ 0 & \text{otherwise} \end{cases}$$

$$(5.11)$$

Now, all $x_{(u,v,w)}^s$ variables having $u = v = w$ can be interpreted as registers and are weighted with $\mathrm{cost_R}$, all others require one $\mathcal{A}$-operation and are weighted with $\mathrm{cost_A}$. Now, no additional variables are necessary to distinguish between registers and a single binary variable x_w^s is used which is '1' when w is realized in stage s. The alternative PMCM ILP formulation is given in PMCM ILP Formulation 2.

Constraints C1, C2 and C3 correspond to the modified versions of C1, C4 and C5 in PMCM ILP Formulation 1. C1 constraints that the target elements are located in the last stage, C2 and C3 ensures that at least one valid (u,v) pair is present in the preceding stage to compute w. The relaxation $x_{(u,v,w)}^s \in \mathbb{R}_0$ leads to binary solutions $x_{(u,v,w)}^s \in \{0,1\}$ which can be proven in the same way as done for PMCM ILP Formulation 1.

PMCM ILP Formulation 2.

$$\min \sum_{s=1}^{S} \sum_{(u,v,w)\in\mathcal{T}^s} \text{cost}(u,v,w)\, x_{(u,v,w)}^s$$

with

$$\text{cost}(u,v,w) = \begin{cases} \text{cost}_R(w) & \text{for } u = v = w \\ \text{cost}_A(u,v,w) & \text{otherwise} \end{cases}$$

subject to

C1: $\quad x_w^S = 1$ for all $w \in T_{\text{odd}}$

C2: $\quad x_w^s - \displaystyle\sum_{(u,v,w')\in\mathcal{T}^s \,|\, w'=w} x_{(u,v,w)}^s = 0$ for all $w \in \mathcal{S}^s, s = 1 \ldots S$

C3: $\quad \left.\begin{aligned} x_{(u,v,w)}^s - x_u^{s-1} \leq 0 \\ x_{(u,v,w)}^s - x_v^{s-1} \leq 0 \end{aligned}\right\}$ for all $(u,v,w) \in \mathcal{T}^s$, $s = 2 \ldots S$

$\quad x_w^s \in \{0,1\}$ for all $w \in \mathcal{S}^s$, $s = 1 \ldots S$

$\quad x_{(u,v,w)}^s \in \mathbb{R}_0$ for all $(u,v,w) \in \mathcal{T}^s, s = 1 \ldots S$

Note that the constraint C3 in PMCM ILP Formulation 2

$$x_{(u,v,w)} - x_u \leq 0 \tag{5.12}$$

$$x_{(u,v,w)} - x_v \leq 0 \tag{5.13}$$

is equivalent but more restrictive in the relaxed model than the corresponding constraint C3 in ILP Formulation 3 which is repeated here (without loss of generality the stage information was omitted):

$$x_u + x_v \geq 2x_{(u,v,w)} \tag{5.14}$$

Both constraints require that if $x_{(u,v,w)}$ is '1', then the nodes x_u and x_v have to be '1', too. In addition, both x_u and x_v have to be unconstrained when $x_{(u,v,w)} = 0$. Obviously, this is fulfilled for both constraints making them interchangeable. Introducing a difference variable as $\delta = x_v - x_u$ ($-1 \leq \delta \leq 1$) constraint (5.14) results in

$$x_{(u,v,w)} \leq \frac{1}{2}x_u + \frac{1}{2}x_v = x_u + \frac{1}{2}\delta \tag{5.15}$$

and constraints (5.12) and (5.13) lead to

$$x_{(u,v,w)} \leq x_u \tag{5.16}$$
$$x_{(u,v,w)} \leq x_v = x_u + \delta . \tag{5.17}$$

Clearly, if $\delta = 0$ all constraints result in the same limit $x_{(u,v,w)} \leq x_u$, if $\delta > 0$, (5.16) is tighter than (5.15) and if $\delta < 0$ (5.17) is tighter than (5.15). This tighter bound typically leads to a faster optimization.

5.2.3 PMCM ILP Formulation 3

The third and last ILP formulation is based on the fact that rearranging constraint C2 in PMCM ILP Formulation 2 leads to a fixed relation of variable x_w^s:

$$x_w^s = \sum_{(u,v,w') \in \mathcal{T}^s \,|\, w'=w} x_{(u,v,w)}^s . \tag{5.18}$$

This can be substituted into constraints C1 and C3 of PMCM ILP Formulation 2 while eliminating C2, leading to PMCM ILP Formulation 3.

The formulation involves less (but pure binary) variables and less constraints (but including more variables).

5.3 Extensions to Adder Depth Constraints and Glitch Path Count Minimization

The PMCM formulations of the last section can easily be modified to the $\mathrm{MCM_{BAD}}$ and $\mathrm{MCM_{MAD}}$ problems (see Section 2.4.5 for their definitions). For that, pipeline registers are replaced by wires with no cost by setting $\mathrm{cost_R}(w) = 0$ for both problems.

PMCM ILP Formulation 3.

$$\min \sum_{s=1}^{S} \sum_{(u,v,w) \in \mathcal{T}^s} \text{cost}(u,v,w)\, x^s_{(u,v,w)}$$

with

$$\text{cost}(u,v,w) = \begin{cases} \text{cost}_R(w) & \text{for } u = v = w \\ \text{cost}_A(u,v,w) & \text{otherwise} \end{cases}$$

subject to

C1:
$$\sum_{(u,v,w') \in \mathcal{T}^s \,|\, w'=w} x^S_{(u,v,w)} = 1 \text{ for all } w \in T_{\text{odd}}$$

C2:
$$\left.\begin{aligned} x^s_{(u,v,w)} - \sum_{(u',v',w') \in \mathcal{T}^s \,|\, w'=u} x^{s-1}_{(u',v',w)} \leq 0 \\ x^s_{(u,v,w)} - \sum_{(u',v',w') \in \mathcal{T}^s \,|\, w'=v} x^{s-1}_{(u',v',w)} \leq 0 \end{aligned}\right\} \begin{aligned} &\text{for all } (u,v,w) \in \mathcal{T}^s \\ &\text{with } s = 2 \ldots S \end{aligned}$$

$$x^s_{(u,v,w)} \in \{0,1\} \text{ for all } (u,v,w) \in \mathcal{T}^s$$
$$\text{with } s = 1 \ldots S$$

While this is already sufficient for the MCM_{BAD} problem, the constraint C1 in ILP formulation 1, 2 and 3 has to be modified for the MCM_{MAD} problem to

$$r^s_w + a^s_w = 1 \text{ for all } w \in T_{\text{odd}} \text{ with } s = \text{AD}(w) \,, \quad (5.19)$$

$$x^s_w = 1 \text{ for all } w \in T_{\text{odd}} \text{ with } s = \text{AD}(w) \quad (5.20)$$

and

$$\sum_{(u,v,w') \in \mathcal{T}^s \,|\, w'=w} x^s_{(u,v,w)} = 1 \text{ for all } w \in T_{\text{odd}} \text{ with } s = \text{AD}(w) \,, \quad (5.21)$$

respectively.

Compared to ILP Formulation 3, which can also be used for MCM_{MAD}, no depth variables or MTZ constraints are required. This comes along with a slight increase of variables representing nodes in the corresponding hypertree due to the duplication of nodes in several stages and their differentiation in adder or register realizations. Furthermore, using PMCM ILP Formulation 1 uses much fewer (u, v) pairs than the (u, v, w) triplets in ILP Formulations 3 or 2.

Another extension is the direct optimization of the glitch path count (GPC), which gives a high level estimate for the dynamic power [50]. So far, the GPC was reduced by optimizing MCM for low or minimal AD but there may be different solutions which provide minimal AD but a different GPC. The GPC of an adder node is simply computed by

$$\text{GPC}_{\text{out}} = \text{GPC}_{\text{in},1} + \text{GPC}_{\text{in},2} + 1 \tag{5.22}$$

where $\text{GPC}_{\text{in},1}$ and $\text{GPC}_{\text{in},2}$ are the GPC values of the input node. The GPC of the input of the MCM block is defined to be zero. The problem of the inclusion of the GPC in the ILP model is that there is no 1:1 relation of a node variable and its GPC. This would lead to constraints like $\text{GPC}_w^s = \text{GPC}_u^s x_{(u,v,w)}^s + \text{GPC}_v^s x_{(u,v,w)}^s + 1$ which is non-linear as both GPC_u^s and $x_{(u,v,w)}^s$ have to be variable. However, the following constraints can be added to PMCM formulations 2 or 3 to include the GPC:

$$-L + 1 \le \text{GPC}_w^s - \text{GPC}_u^{s-1} - \text{GPC}_v^{s-1} - L x_{(u,v,w)}^s \tag{5.23}$$

$$\text{GPC}_w^s - \text{GPC}_u^{s-1} - \text{GPC}_v^{s-1} + L x_{(u,v,w)}^s \le L + 1 \tag{5.24}$$

for all $(u, v, w) \in \mathcal{T}^s, s = 1 \dots S$ with $w \ne u$, $w \ne v$. The integer variable $\text{GPC}_w^s \in \mathbb{Z}_0$ is required to each single element in each stage ($w \in \mathcal{S}^s$, $s = 0 \dots S$) with $\text{GPC}_1^0 := 0$.

For $x_{(u,v,w)}^s = 0$, the constraints (5.23) and (5.24) lead to

$$-L \le \text{GPC}_w^s - \text{GPC}_u^{s-1} - \text{GPC}_v^{s-1} - 1 \le L , \tag{5.25}$$

which does not constraint GPC_w^s for sufficiently large L (more precisely, constant L has to be chosen to be larger than the maximum possible GPC). With $x_{(u,v,w)}^s = 1$ the constraints result in

$$1 \le \text{GPC}_w^s - \text{GPC}_u^{s-1} - \text{GPC}_v^{s-1} \le 1 \tag{5.26}$$

leading to

$$\text{GPC}_w^s = \text{GPC}_u^{s-1} + \text{GPC}_v^{s-1} + 1 \tag{5.27}$$

which is the definition of the GPC. Now, the objective can be changed to:

$$\min \sum_{s=1}^{S} \sum_{w \in \mathcal{S}^s} \text{GPC}_w^s \qquad (5.28)$$

leading to a minimal GPC formulation.

5.4 Further Extensions of the ILP Formulations

The proposed ILP formulations are very generic and extendable into different directions. The cost model of PMCM formulations 2 and 3 can be easily adapted to ASIC low-level cost models by exchanging the cost functions [45–47].

Like the extended ILP formulation for the optimal pipelining of precomputed adder graphs in Section 3.2.2, modified cost functions can also be used to chose a trade-off between latency and throughput. The latency (and FF resources) can be reduced at the cost of throughput by placing pipeline registers every n'th stage (e. g., every second or third stage) instead of placing pipeline registers in every stage. To do so, the cost function for pure registers can be defined to be zero in each stage in which no registers are designated. These registers are exchanged by wires in the implementation. Reversely, the throughput can be increased at the cost an increased latency (and resources) by using pipelined adders. For that, the cost functions for adders and registers have to be increased according to the resource consumption of the pipelined adder and multiple register stages, respectively.

Another extension would be the support of dedicated shift registers which are provided by the latest Xilinx FPGA architectures. These support shift registers of up to 16 bit on Virtex 4 FPGAs (SRL16 primitive) and up to 32 bit on Virtex 5/6, Spartan 6 and 7 series FPGAs (SRLC32E primitive) and are realized in a single FPGA LUT. Thus, a single BLE can realize up to 16 or 32 registers (plus one additional register at the output). As long as the pipeline depth is $S \leq 17$ (Virtex 4 FPGAs) or $S \leq 33$ (Virtex 5/6/7

FPGAs), constraint (C3) in PMCM ILP Formulation 2 can be replaced by:

$$
\text{(C3a)} \quad
\left.
\begin{aligned}
x^s_{(u,v,w)} - x^{s-1}_u &\leq 0 \\
x^s_{(u,v,w)} - x^{s-1}_v &\leq 0
\end{aligned}
\right\}
\quad
\begin{aligned}
&\text{for all } (u,v,w) \in \mathcal{T}^s,\, s = 2 \ldots S \\
&\text{with } w \neq u,\, w \neq v
\end{aligned}
$$

$$
\text{(C3b)} \quad
x^s_{(u,v,w)} - \sum_{s'=0}^{s-1} x^{s'}_u \leq 0
\quad
\begin{aligned}
&\text{for all } (u,v,w) \in \mathcal{T}^s,\, s = 2 \ldots S \\
&\text{with } w = v \ \vee \ w = u
\end{aligned}
$$

While (C3a) is nearly identical to (C3) but excludes registers, (C3b) allows the bypass of several stages by using shift registers. Note that these values are not accessible for intermediate stages.

5.5 Analysis of Problem Sizes

5.5.1 Evaluation of Set Sizes of $\mathcal{A}^s$ and $\mathcal{T}^s$

The number of variables in the ILP formulations above is directly related to the number of elements in $\mathcal{A}^s$ and $\mathcal{T}^s$, which correspond to the number of possible odd fundamentals and $\mathcal{A}$-configurations to realize them in stage s, respectively. The set sizes were evaluated numerically for different values of the target coefficient word size B_T and pipeline stages s and are listed in Table 5.1. Missing values which were computationally too complex are marked with '−'. Note that it was assumed that all possible elements with up to B_T bit were required in stage s, (i. e., $\mathcal{S}^s = \mathcal{A}^s$) to compute $\mathcal{T}^s$. As the higher stages are not needed for lower bit width coefficients, only the bold entries are relevant.

5.5.2 Relation to FIR Filters

Taking the FIR filter as one of the most important MCM application, the coefficient word size is directly related to the approximation quality. Early studies about this impact have shown that a coefficient word sizes between 15 bit to 20 bit are sufficient for approximation errors between -70 and -100 dB [18]. Hence, for most practical FIR filters an adder depth of three should be sufficient while only an adder depth of four may be necessary for high performance filters.

Table 5.1: Number of elements in $\mathcal{A}^s$ and upper bound of triplets in $\mathcal{T}^s$ for pipeline stages $s = 1 \ldots 4$ and bit width $B_T = 2 \ldots 32$

B_T	$|\mathcal{A}^1|$ / $|\mathcal{T}^1|$	$|\mathcal{A}^2|$ / $|\mathcal{T}^2|$	$|\mathcal{A}^3|$ / $|\mathcal{T}^3|$	$|\mathcal{A}^4|$ / $|\mathcal{T}^4|$
2	2 / 2	2 / 4	2 / 4	2 / 4
4	6 / 6	8 / 69	8 / 123	8 / 123
6	10 / 10	32 / 379	32 / 2,744	32 / 2,744
8	14 / 14	120 / 1,144	128 / 43,348	128 / 47,779
10	18 / 18	336 / 2,549	512 / $403 \cdot 10^3$	512 / 780,881
12	22 / 22	744 / 4,786	2,048 / $2 \cdot 10^6$	2,048 / $13 \cdot 10^6$
14	26 / 26	1,408 / 8,047	8,192 / $10 \cdot 10^6$	8,192 / $201 \cdot 10^6$
16	30 / 30	2,392 / 12,524	32,640 / $35 \cdot 10^6$	32,768 / $3 \cdot 10^9$
18	34 / 34	3,760 / 18,409	$124 \cdot 10^3$ / $99 \cdot 10^6$	$131 \cdot 10^3$ / $47 \cdot 10^9$
20	38 / 38	5,576 / 25,894	$423 \cdot 10^3$ / $248 \cdot 10^6$	$524 \cdot 10^3$ / $580 \cdot 10^9$
22	42 / 42	7,904 / 35,171	$1 \cdot 10^6$ / $559 \cdot 10^6$	– / –
24	46 / 46	10,808 / 46,432	$3 \cdot 10^6$ / $1 \cdot 10^9$	– / –
26	50 / 50	14,352 / 59,869	$8 \cdot 10^6$ / $2 \cdot 10^9$	– / –
28	54 / 54	18,600 / 75,674	$17 \cdot 10^6$ / $4 \cdot 10^9$	– / –
30	58 / 58	23,616 / 94,039	$35 \cdot 10^6$ / $7 \cdot 10^9$	– / –
32	62 / 62	29,464 / $115 \cdot 10^3$	$68 \cdot 10^6$ / $12 \cdot 10^9$	– / –

5.6 Experimental Results

Different experiments were performed to answer the following questions:

1. Which is the best (fastest) ILP formulation for the PMCM problem?

2. How does the runtime scale and what are feasible problem sizes?

3. How good is the quality of the RPAG heuristic compared to solutions which are known to be optimal?

For that, a C++ tool called optimal pipelined adder graph (OPAG) was implemented that provides a similar command line interface as the RPAG implementation. It takes the coefficients and calls the library of an MILP solver. The MILP solver interface was encapsulated using the Google or-tools library which provides a unique wrapper for most state-of-the-art open-source and commercial ILP/MILP solvers [144]. The solver used in the OPAG implementation can be selected by setting --ilp_solver to scip [129], cplex [145] or gurobi [146]. As a side effect, different MILP optimizers were evaluated for their suitability for the used ILP formulations. Different optimization problems are implemented which can be selected by setting --problem to pmcm, mcmmad or mcmmgpc for the PMCM the MCM_{MAD} or the minimum GPC problem, respectively. Its implementation is freely available as open-source [16].

5.6.1 FIR Filter Benchmark

In the first experiment, the same filters as used in Section 3.3.1 and Section 4.5.6 were used, originally published by Mirzaei et al. [5]. Due to the large problem size, the ILP solver (CPLEX [145]) could only complete the first two instances of the benchmark. For the other filters, the memory requirements exceeded the available memory of 48 GB. The RPAG solution could be improved by one registered operation, from 9 to 8 and from 12 to 11 registered operation for $N = 6$ and $N = 10$, respectively. The resulting PAGs are compared with the PAGs obtained by RPAG in Figure 5.2.

From these examples, the limitations of the heuristic can be observed. For the $N = 6$ filter, RPAG is able to find a predecessor set with three elements from which all target elements can be computed. The same number of elements is used in stage two of the optimal PAG. However, the elements $\{5, 37, 8159\}$ in the optimal PAG can be realized with only two predecessors which is not possible for the elements $\{1, 165, 10207\}$ used in the RPAG solution (this was proven by optimally solving the PMCM problem for these coefficients). Exactly the same behavior can be observed in the $N = 10$ filter. While $\{255, 51, 61, 125\}$ can be computed from two predecessors, at least three predecessors are necessary to compute $\{1, 25, 189, 503\}$. It is noticeable that coefficient '1' is computed from predecessors by using right shifts in both optimal results. Right shifts are also considered in RPAG but they are not used in this case as the predecessor '1' (to compute working set element '1') offers a gain of one (like all other single predecessors in both examples) using the high-level cost model while having the lowest magnitude, leading to its early selection.

To provide more benchmark results, coefficient sets from image processing applications with small word size were taken as benchmark for remaining experiments. These are small enough that the optimal method can be applied but still useful for realistic applications. They were already used in a previous publication [139] and are briefly introduced in the following.

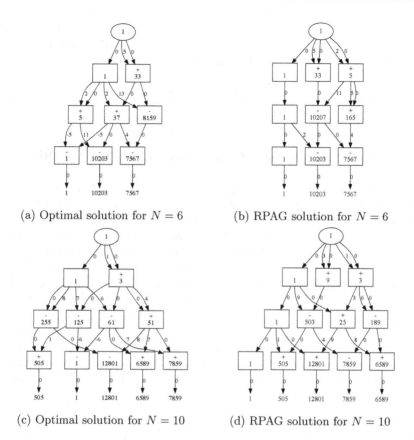

(a) Optimal solution for $N = 6$ (b) RPAG solution for $N = 6$

(c) Optimal solution for $N = 10$ (d) RPAG solution for $N = 10$

Figure 5.2: PAG solutions of two instances of the MIRZAEI10 benchmark

5.6.2 Benchmark Filters from Image Processing

Due to the low coefficient word size of typically 8 to 12 bit in the domain of image processing, a short convergence time of the ILP solver is very likely, which makes the proposed optimization an ideal candidate for image processing. Hence, a benchmark set of 2D folding matrices was designed using Matlab and PMCM ILP Formulation 2 was used which allows a bit-level cost model. If one sample is processed each clock cycle, a common MCM circuit is required that provides the multiplication with each matrix element [139].

The coefficients used in the benchmark were computed using the Matlab fspecial() function except for the lowpass and highpass filters. These were obtained by the Parks-McClellan algorithm (`remez()` and `ftrans2()` functions). The obtained real coefficients were quantized to a word size of B_c. The complete convolution matrices are given in Appendix A.1, the filter parameters are summarized in Table 5.2 with their matrix size, word size, the required pipeline stages S, parameters for their design and the N_{uq} unique odd coefficients of their folding matrix.

5.6.3 Optimization Results with High-Level Cost Model

In the second experiment, the high-level cost model is used where the goal is to find the PAG with least number of registered operations. Results were obtained from the presented ILP formulations and, as reference, the original RPAG algorithm, denoted as RPAG (v1) [8] and the improved RPAG (v2) implementation of Chapter 4. RPAG (v1) was performed with a single run ($R = 1$) and 50 runs ($R = 50$), were predecessors with identical gain are randomly selected and the best result is taken. The search limit of RPAG (v2) was set to $L = 1000$, which was practically limited by the number of available predecessors. The results are tabulated in Table 5.3. The timeout (TO) limit of 8 hours was exceeded for the lowpass 15 × 15 instance, here, the best obtained result is shown as upper bound. It can be observed that the results of the RPAG (v2) heuristic are always very close or identical to the optimal solution. However, an average reduction by 3.2% compared to RPAG (v2) can be achieved with the optimal method.

The optimal results were obtained using three different ILP optimizers, namely the commercial optimizers CPLEX (v12.06) [145] and Gurobi (v6.0.0) [146] as well as the open-source optimizer SCIP (v3.1.0) [129]. Their computation times on an Intel Nehalem 2.93 GHz computer with 48 GB memory are listed in Table 5.4. While an optimal solution was found for all depth-two ($S = 2$) coefficients within less than one second, the computation time was much longer for the three instances with $S = 3$. For some of these cases, the timeout of 8 hours was exceeded. Due to different optimization strategies of the different optimizers, there is no general answer to the question which is the best ILP formulation. Obviously, the shortest computation time was obtained by the state-of-the-art optimizer Gurobi using PMCM ILP Formulation 1. Gurobi also found the best solution for the lowpass 15 × 15 filter instance, where the timeout of all optimizers was reached. While Gurobi obtained an approximate solution with 32 registered operations after eight hours, the best solutions of CPLEX and SCIP were 34 and 35 registered operations, respectively. In contrast to that, the computation

Table 5.2: Parameters of the used benchmark filters

Filter Type	Filter Size	B_c	N_{uq}	S	Parameter	Unique odd coefficients
gaussian	3 × 3	8	3	2	$\sigma = 0.5$	3 21 159
gaussian	5 × 5	12	4	3		1 23 343 1267
laplacian	3 × 3	8	3	2	$\alpha = 0.2$	5 21 107
unsharp	3 × 3	8	3	2	$\alpha = 0.2$	3 11 69
unsharp	3 × 3	12	3	3		43 171 1109
lowpass	5 × 5	8	5	2	$f_{pass} = 0.2$, $f_{stop} = 0.4$	11 33 35 53 103
lowpass	9 × 9	10	13	2		1 5 7 25 31 63 65 67 73 97 117 165 303
lowpass	15 × 15	12	26	3		1 5 7 13 17 19 21 27 41 43 45 53 61 79 93 101 103 113 133 137 199 331 333 613 1097 1197
highpass	5 × 5	8	5	2	$f_{stop} = 0$, $f_{pass} = 0.2$	1 3 5 7 121
highpass	9 × 9	10	6	2		1 3 5 7 11 125
highpass	15 × 15	12	13	2		1 3 5 7 9 11 13 15 17 19 21 23 507

time of RPAG (v2) was always less than four seconds.

5.6.4 Optimization Results with Low-Level Cost Model

In the third experiment, results using the low-level cost model were obtained. All PMCM blocks were optimized for an input word size B_i of 8, 10 and 12 bit. The results are summarized in Table 5.5 for RPAG ($R = 1$ and $R = 50$) and RPAG (v2) which are compared to the optimal pipelined adder graph method. It can be seen that for the low-level cost model, the RPAG solutions are closer to the optimum than for the high-level model. An average improvement of less that 1% can be obtained by using the optimal ILP-based method.

5.6.5 Synthesis Results

As reality check for the low-level cost model, the optimal designs of the last section were synthesized in the fourth experiment. For that, a VHDL code generator was developed in Matlab. The synthesis was performed for a

Table 5.3: Optimization results for the high-level cost model in terms of the number of registered operations for RPAG (v1) [8] using R iterations, RPAG (v2) and the optimal pipelined adder graphs (best marked bold)

Filter Type	Filter Size	B_c	registered operations			
			RPAG (v1) [8] ($R = 1$)	RPAG (v1) [8] ($R = 50$)	RPAG (v2)	Optimal PAG
gaussian	3×3	8	**5**	**5**	**5**	**5**
gaussian	5×5	12	10	**9**	**9**	**9**
laplacian	3×3	8	7	**5**	**5**	**5**
unsharp	3×3	8	**5**	**5**	**5**	**5**
unsharp	3×3	12	10	9	9	**7**
lowpass	5×5	8	**8**	**8**	**8**	**8**
lowpass	9×9	10	18	**17**	**17**	**17**
lowpass	15×15	12	35	34	34	**≤32**
highpass	5×5	8	**7**	**7**	**7**	**7**
highpass	9×9	10	**8**	**8**	**8**	**8**
highpass	15×15	12	**16**	**16**	**16**	**16**
Avg.:			11.73	11.18	11.18	10.82
Imp. to RPAG (v1) $R = 1$:			–	4.65%	4.65%	7.75%

Virtex 6 XC6VLX75T-2FF484 FPGA using Xilinx ISE v13.4. One problem in the evaluation of the low-level cost model was the correct counting of BLEs. The synthesis report contains the number of FFs as well as the number of LUTs and their type (O5, O6), but some BLEs use only the LUT, only the FFs or both, which makes it impossible to compute the correct number of BLEs from the synthesis report. Therefore, a small tool was written that counts the BLEs from the netlist. The tool parses the Xilinx design language (XDL) representation of the netlist by using the XDL parser from the RapidSmith Project [147] and analyzes the LUTs and FFs according to their location to obtain the correct number of BLEs.

The BLE counts from the low-level model and from synthesis as well as the maximum clock frequency $f_{\max}$ obtained by `trace` are listed in Table 5.6. It can be seen that the low-level model precisely predicts the synthesis in most cases. However, there are some cases where the actual costs are lower than predicted. Analyzing the designs using the Xilinx FPGA editor revealed the following reasons:

1. A single Virtex 6 BLE is able to realize up to two FFs (see Fig-

Table 5.4: Runtime comparison of different ILP formulations of the PMCM problem with high level cost function using different solvers and a timeout (TO) of 8 h/28,800 sec

	runtime [sec]								
Solver:	CPLEX			Gurobi			SCIP		
PMCM ILP Formulation:	1	2	3	1	2	3	1	2	3
gaussian 3 × 3 8 bit	< 1	< 1	< 1	< 1	< 1	< 1	< 1	0.018	< 1
gaussian 5 × 5 12 bit	10,232	23,422	2,590	1,758	8,944	TO	26,292	TO	7,987
laplacian 3 × 3 8 bit	< 1	< 1	< 1	< 1	< 1	< 1	< 1	< 1	< 1
unsharp 3 × 3 8 bit	< 1	< 1	< 1	< 1	< 1	< 1	< 1	< 1	< 1
unsharp 3 × 3 12 bit	1,301	4,743	1,475	849	1,947	2,876	9,020	TO	2,132
lowpass 5 × 5 8 bit	< 1	< 1	< 1	< 1	< 1	< 1	< 1	< 1	< 1
lowpass 9 × 9 10 bit	< 1	< 1	< 1	< 1	< 1	< 1	< 1	< 1	< 1
lowpass 15 × 15 12 bit	TO	TO	TO	TO	TO	TO	TO	TO	TO
highpass 5 × 5 8 bit	< 1	< 1	< 1	< 1	< 1	< 1	< 1	< 1	< 1
highpass 9 × 9 10 bit	< 1	< 1	< 1	< 1	< 1	< 1	< 1	< 1	< 1
highpass 15 × 15 12 bit	< 1	< 1	< 1	< 1	< 1	< 1	< 1	< 1	< 1

ure 2.6(b)) but most of the times, only one FF is used by ISE. This was modeled by setting $K_{FF} = 1$ in (2.27) (see page 27) leading to an overestimation as sometimes two FFs are realized in the same BLE.

2. Sometimes ISE maps a single FF into a BLE that is also used for a full adder (FA).

Hence, the cost model sometimes overestimates the real costs by assuming that each single FF is mapped to a single BLE. It is noticeable that the speed of the PAG is in the same order of magnitude as the embedded DSP48E1 block which can be clocked with 600 MHz. In 75.8% of the cases using the optimal PAG method, the resulting circuit is even faster than a DSP block.

5.6.6 Optimization Results for Minimal Adder Depth

In the last experiment, the modifications for solving the MCM_{MAD} problem as discussed in Section 5.3 were evaluated. For that, the benchmark of Aksoy et al. [9] is used which was also used for the minLD algorithm [34]. The results are listed in Table 5.7. The numeric numbers of the filters are available online [127] as AKSOY08_NORCHIP_ID where ID is the filter ID in Table 5.7. The results from minLD [34] are listed as reference. In

Table 5.5: Optimization results for the low-level cost model in terms of the number of BLEs for RPAG (v1) [8] using R iterations, RPAG (v2) and the optimal pipelined adder graphs (best marked bold)

Filter Type	Filter Size	B_c	B_i	RPAG (v1) [8] ($R = 1$)	RPAG (v1) [8] ($R = 50$)	RPAG (v2)	Optimal PAG
gaussian	3 × 3	8	8	63	**58**	**58**	**58**
gaussian	5 × 5	12	8	125	125	**111**	**111**
laplacian	3 × 3	8	8	79	**61**	**61**	**61**
unsharp	3 × 3	8	8	**56**	**56**	**56**	**56**
unsharp	3 × 3	12	8	112	107	107	**91**
lowpass	5 × 5	8	8	**98**	**98**	**98**	**98**
lowpass	9 × 9	10	8	235	**221**	**221**	**221**
lowpass	15 × 15	12	8	480	475	470	**≤478**
highpass	5 × 5	8	8	**74**	**74**	**74**	**74**
highpass	9 × 9	10	8	**85**	**85**	**85**	**85**
highpass	15 × 15	12	8	**186**	**186**	**186**	**186**
gaussian	3 × 3	8	10	73	**68**	**68**	**68**
gaussian	5 × 5	12	10	145	143	**129**	**129**
laplacian	3 × 3	8	10	93	**71**	**71**	**71**
unsharp	3 × 3	8	10	**66**	**66**	**66**	**66**
unsharp	3 × 3	12	10	130	123	123	**105**
lowpass	5 × 5	8	10	**114**	**114**	**114**	**114**
lowpass	9 × 9	10	10	271	**255**	**255**	**255**
lowpass	15 × 15	12	10	550	**543**	538	**≤547**
highpass	5 × 5	8	10	**88**	**88**	**88**	**88**
highpass	9 × 9	10	10	**101**	**101**	**101**	**101**
highpass	15 × 15	12	10	**218**	**218**	**218**	**218**
gaussian	3 × 3	8	12	83	**78**	**78**	**78**
gaussian	5 × 5	12	12	165	161	**147**	**147**
laplacian	3 × 3	8	12	107	**81**	**81**	**81**
unsharp	3 × 3	8	12	**76**	**76**	**76**	**76**
unsharp	3 × 3	12	12	148	139	139	**119**
lowpass	5 × 5	8	12	**130**	**130**	**130**	**130**
lowpass	9 × 9	10	12	307	**289**	**289**	**289**
lowpass	15 × 15	12	12	620	611	605	**≤620**
highpass	5 × 5	8	12	**102**	**102**	**102**	**102**
highpass	9 × 9	10	12	**117**	**117**	**117**	**117**
highpass	15 × 15	12	12	**250**	**250**	**250**	**250**
			Average:	168.09	162.73	160.97	160.30
	Improvement to RPAG (v1) $R = 1$:			–	3.19%	4.24%	4.63%

Table 5.6: Comparison between synthesis results using the same benchmark instances as in Table 7.1 providing the actual BLEs as well as the maximum clock frequency f_{max} (best marked bold) on a Virtex 6 FPGA

Filter Type	Filter Size	B_c	B_i	No. of BLE LL model	No. of BLE synthesis	Rel. BLE Error	f_{max} [MHz]
gaussian	3×3	8	8	58	58	0%	730.5
gaussian	5×5	12	8	111	111	0%	619.2
laplacian	3×3	8	8	61	61	0%	701.8
unsharp	3×3	8	8	56	56	0%	709.7
unsharp	3×3	12	8	91	91	0%	657.9
lowpass	5×5	8	8	98	98	0%	657.9
lowpass	9×9	10	8	221	221	0%	582.4
lowpass	15×15	12	8	478	478	0%	489.0
highpass	5×5	8	8	74	74	0%	711.7
highpass	9×9	10	8	85	85	0%	689.2
highpass	15×15	12	8	186	183	1.61%	598.4
gaussian	3×3	8	10	68	68	0%	745.7
gaussian	5×5	12	10	129	129	0%	637.8
laplacian	3×3	8	10	71	71	0%	656.6
unsharp	3×3	8	10	66	66	0%	712.3
unsharp	3×3	12	10	105	103	1.90%	605.3
lowpass	5×5	8	10	114	114	0%	646.8
lowpass	9×9	10	10	255	251	1.57%	602.1
lowpass	15×15	12	10	547	546	0.18%	514.7
highpass	5×5	8	10	88	86	2.27%	704.2
highpass	9×9	10	10	101	97	3.96%	677.5
highpass	15×15	12	10	218	218	0%	567.9
gaussian	3×3	8	12	78	78	0%	644.3
gaussian	5×5	12	12	147	147	0%	624.2
laplacian	3×3	8	12	81	81	0%	724.1
unsharp	3×3	8	12	76	72	5.26%	638.6
unsharp	3×3	12	12	119	119	0%	638.6
lowpass	5×5	8	12	130	126	3.08%	653.2
lowpass	9×9	10	12	289	278	3.81%	581.7
lowpass	15×15	12	12	620	619	0.16%	530.8
highpass	5×5	8	12	102	94	7.84%	714.3
highpass	9×9	10	12	117	113	3.42%	680.7
highpass	15×15	12	12	250	239	4.40%	539.7

addition, the results from RPAG with the MCM_{MAD} extensions as described in Section 4.1.2 are provided. It can be observed that the solutions provided by minLD and RPAG are very close to the optimum in terms of adder count.

Table 5.7: MCM$_{MAD}$ benchmark with filters from Aksoy [9]

Filter ID	AD$_{min}$	minLD [34] #Add	RPAG (MCM$_{MAD}$) #Add	optimal min. AD #Add	CPU time
A40	3	**26**	26	**26**	1.2 s
B80	3	**51**	52	$\leq$ **51**	$\geq$ 36 h
C30	3	**18**	18	**18**	3 s
D80	3	**38**	39	**38**	53.4 s
E40	3	**25**	25	**25**	1.2 s
F80	3	**44**	44	**44**	0.6 s
G40	3	**24**	24	**24**	2.1 s
H60	3	38	38	**37**	56.9 min
I40	3	**25**	25	**25**	0.4 s
J60	3	38	38	**37**	20.1 min
K25	3	**19**	20	**19**	257 s
avg.:		31.45	31.73	31.27	

Interestingly, the MCM$_{MAD}$ problem can be solved much faster by ILP solvers than the PMCM problem. Most of the instances listed in Table 5.7 which only took seconds for MCM$_{MAD}$ resulted in a timeout (8h) when solved for the PMCM problem (using the same optimizer, Gurobi, and the same base ILP formulation). Setting the register cost to zero (like for the MCM$_{BAD}$ problem) yields to similar large optimization times as for the PMCM problem. Apparently, the fact that elements in lower stages are already predetermined in the MCM$_{MAD}$ problem drastically simplifies the optimization. It seems to be plausible that this reduces the search space as the selection of non-output fundamentals which can be beneficially combined with already fixed elements is more restricted.

5.7 Conclusion

Different ILP formulations to optimally solve the PMCM and related problems were presented in this section. Even though ILP problems are known to be NP-hard, it was shown that problem sizes of practical interest can be solved successfully. In addition, the quality of the results from the RPAG heuristic as presented in Chapter 4 was evaluated using the known optimal solutions. In most of the cases, the optimal solution or a solution very close to the optimum was found. The RPAG results were closer to the optimal value when the low-level cost model was used. There, the gap to the optimal solution was only 0.42% on average for low word-size coefficients.

6 A Heuristic for the Constant Matrix Multiplication Problem

This chapter deals with the problem of optimizing the multiplication of a constant matrix with a vector, the constant matrix multiplication (CMM) operation in a pipelined fashion (for an introduction of the CMM problem, see Section 2.3). Finding a PAG with minimum adder count for this kind of problem, in the following denoted as the PCMM problem, is a generalization of the PMCM problem. As the search space of the PCMM problem is much larger compared to the PMCM problem, the aim is to heuristically solve the problem by extending the RPAG algorithm of Chapter 4. This extension is available in the open-source RPAG implementation (command line flag --cmm) [16]. Like the PMCM problem, the PCMM problem can be easily adjusted to solve the CMM problem with bounded or minimal AD, denoted as CMM_{BAD} and CMM_{MAD} in the following. As none of the available CMM algorithms support AD constraints, this might also be useful for low-power ASIC designs.

The CMM operation of an $M \times N$ matrix $\mathbf{C}$ with vector $\vec{x}$ is represented in the following form:

$$\begin{pmatrix} y_1 \\ y_2 \\ \vdots \\ y_M \end{pmatrix} = \begin{pmatrix} c_{1,1} & c_{1,2} & \cdots & c_{1,N} \\ c_{2,1} & c_{2,2} & \cdots & c_{2,N} \\ \vdots & \vdots & \ddots & \vdots \\ c_{M,1} & c_{M,2} & \cdots & c_{M,N} \end{pmatrix} \cdot \begin{pmatrix} x_1 \\ x_2 \\ \vdots \\ x_N \end{pmatrix} \tag{6.1}$$

Each output is computed by an SOP operation from the input vector:

$$y_1 = c_{1,1}x_1 + c_{1,2}x_2 + \ldots + c_{1,N}x_N \tag{6.2}$$

$$y_2 = c_{2,1}x_1 + c_{2,2}x_2 + \ldots + c_{2,N}x_N \tag{6.3}$$

$$\vdots$$

$$y_M = c_{M,1}x_1 + c_{M,2}x_2 + \ldots + c_{M,N}x_N \tag{6.4}$$

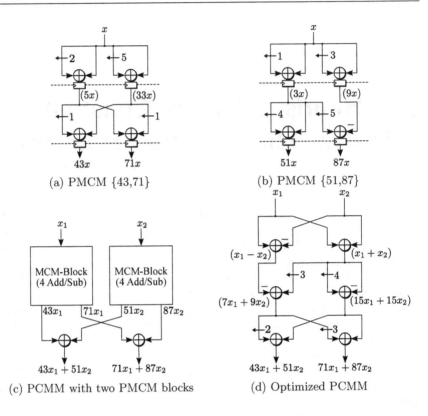

(a) PMCM {43,71}   (b) PMCM {51,87}

(c) PCMM with two PMCM blocks   (d) Optimized PCMM

Figure 6.1: Different realizations of the example PCMM circuit

Hence, it can be realized using M independent SOP operations, each computing one row, or by using N MCM operations, each computing the products of a column, followed by an adder tree. As pointed out in Section 1.4, SOP and MCM are directly related by the transposition, i. e., transposing the single input N-output adder graph of an MCM operation results in an N-input single output SOP operation which processes the same coefficients [20]. The same transformation can be applied to a PMCM operation but this may increase the critical path as registers are placed at different positions. Hence, solving the PCMM problem also provides a solution to optimize pipelined SOP (PSOP) circuits.

To give a first impression about the potential of the CMM optimization, the matrix

$$\mathbf{C} = \begin{pmatrix} 43 & 51 \\ 71 & 87 \end{pmatrix} \qquad (6.5)$$

is used as running example throughout this chapter. One way to realize it is to use two PMCM circuits, each computing the multiples of a matrix column. The corresponding optimal PAGs (obtained by OPAG) are shown in figures 6.1(a) and 6.1(b). They can be combined with two extra adders to get the final PCMM circuit as shown in Figure 6.1(c) which requires 10 adders in total. In contrast to that, an optimized PCMM circuit can be found with the proposed algorithm which uses only six add/subtract operations for the same computation as demonstrated in Figure 6.1(d).

6.1 Related Work

The advantage of directly optimizing CMM circuits instead of using an MCM optimization multiple times was first demonstrated by Hartley [33, 148] and later formalized by Dempster et al. [149]. They extended their Bull and Horrocks modified (BHM) MCM optimization algorithm [44] to solve the CMM problem. This inspired many other authors to develop CMM optimization methods. Another CMM heuristic based on an MCM heuristic [88] was proposed by Gustafsson et al. [150]. They transferred the CMM problem to a graph representation in which a minimum spanning tree (MST) has to be found. This MST yields a new matrix with reduced complexity which is used for the next iteration until the matrix only contains ones and zeros. A CMM method based on CSE was introduced by Macleod and Dempster [151]. They represent the CMM instance by using an $i \times j \times k$ matrix which contains the SD value at bit position k for the constant of row i in column j. Then, two-term subexpressions that occur most often are searched and subsequently eliminated. Another method for optimizing CMM using a genetic algorithm (GA) was proposed by Kinane et al. [38, 152]. Permutations from different SD representations are taken to extract valid SOP sub-terms. These are then combined and selected by a GA.

None of the previous methods for solving the CMM problem considers minimal adder depth solutions or pipelining although it was shown that these are important metrics for low-power and high throughput implementations of MCM circuits [8, 34, 39, 50, 53–55, 125, 139, 153].

6.2 RPAG extension to CMM

An extension of the RPAG algorithm of Section 4 to the PCMM problem and the $\mathrm{CMM_{MAD}}/\mathrm{CMM_{BAD}}$ problems is presented in the following. The main idea is to extend the representation of the adder graph as well as the cost function from scalars to vectors, following the idea of previous CMM algorithms [150].

6.2.1 Adder Graph Extensions to CMM

Each node in a CMM adder graph can be represented by a vector of input weights instead of a scalar weight like for MCM adder graphs [150]. This vector of length N corresponds to the sum of weighted inputs where element i corresponds to the (multiplicative) weight of input x_i, e.g., the vector $(43, 51)$ corresponds to $43x_1 + 51x_2$. As a consequence, the inputs are represented by their corresponding unit vectors. The PAG of the PCMM circuit of Figure 6.1(d) is shown in Figure 6.2.

Now, each node performs an extended $\mathcal{A}$-operation (see (2.6)) with a straight forward element-wise vector extension:

$$\mathcal{A}_q(\vec{u}, \vec{v}) = |2^{l_u}\vec{u} + (-1)^{s_v}2^{l_v}\vec{v}|\,2^{-r} \qquad (6.6)$$

Note that a node vector describes the contribution of each input to the scalar value of the node, hence, the $\mathcal{A}$-operation still corresponds to a scalar add/subtract operation. The sets $\mathcal{A}_*(\vec{u}, \vec{v})$ and $\mathcal{A}_*(X)$ (see Section 2.4.3) are defined accordingly. The corresponding limit $c_{\max}$ defines now the maximum magnitude of each element of the vector and is obtained from the matrix element with the maximum word size according to (2.13).

The minimal AD of a node represented by the vector $\vec{c}$, is extended to the sum over the non-zeros of all elements in the vector [74]:

$$\mathrm{AD_{min}}(\vec{c}) = \left\lceil \log_2\left(\sum_{n=1}^{|\vec{c}|} \mathrm{nz}(c_n) \right) \right\rceil \qquad (6.7)$$

Hence, the minimum AD of the total adder graph can be obtained by finding the maximum of the minimum ADs of the rows of the coefficient matrix $\mathbf{C}$. If $\mathbf{C}$ is defined as a column vector (size M) of row vectors (each size N), i. e, $\mathbf{C} = (\vec{c}_1\vec{c}_2\dots\vec{c}_M)^T$ then, the total AD is [74]:

$$\mathrm{AD_{min}}(\mathbf{C}) = \max_{m=1\dots M} \mathrm{AD_{min}}(\vec{c}_m) \,. \qquad (6.8)$$

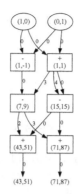

Figure 6.2: Adder graph representation of PCMM circuit in Figure 6.1(d)

Like it was defined for the PMCM problem, the pipeline depth for a PCMM problem is per default set to the minimum possible:

$$S := \mathrm{AD}_{\min}(\mathbf{C}) \tag{6.9}$$

In MCM algorithms, it is beneficial to represent each coefficient by its positive odd fundamental. Hence, is is useful to extend the $\mathrm{odd}(\cdot)$ function to vectors to get a unique representation. For that, all elements are divided by two until at least one element is odd and the sign of all elements is normalized such that the first non-zero element is positive [150]. Note that even elements as well as negative elements may appear in the resulting vector. This normalization can later be reversed by left shifting the result and/or changing the corresponding adders by subtractors (and vice versa).

6.2.2 Definition of the PCMM and $\mathrm{CMM}_{\mathrm{MAD}}/\mathrm{CMM}_{\mathrm{BAD}}$ Problems

Now, the PCMM problem is identical to the PMCM problem (see Definition 7 in Section 4.1) with the target elements in T being vectors (each containing the elements of one matrix row) and pipeline set X_0 is initialized to all possible unit vectors of length N. The same is done to extend the $\mathrm{MCM}_{\mathrm{BAD}}$ and $\mathrm{MCM}_{\mathrm{MAD}}$ problems to the $\mathrm{CMM}_{\mathrm{BAD}}$ and $\mathrm{CMM}_{\mathrm{MAD}}$ problems (see Section 4.1.2).

In the running example, both row vectors of the matrix have an adder depth of three, leading to $X_3 = \{(43, 51), (71, 87)\}$ (for all different CMM problems). Note that for this example, the CMM_{BAD} and CMM_{MAD} problems are identical. As we have two inputs, pipeline set zero is initialized to $X_0 = \{(0,1),(1,0)\}$ and the remaining sets are initialized to empty sets $X_1 = \emptyset$ and $X_2 = \emptyset$. Thus, the minimal cost elements of X_1 and X_2 have to be found by the optimization which is further detailed in the following.

6.2.3 Computation of Predecessors

The vector extension of the computation of single predecessors using topology (a)-(c) as defined in Section 4.2.2 is straight forward by simply replacing the $\mathcal{A}$-operation and AD computation by its vector extensions as defined above:

$$P_a = \{\vec{w} \in W \mid \text{AD}_{\min}(\vec{w}) < s\} \tag{6.10}$$

$$P_b = \{\vec{w}/(2^k \pm 1) \mid \vec{w} \in W, \ k \in \mathbb{N}\} \cap \mathbb{N}^N \setminus \{\vec{w}\} \tag{6.11}$$

$$P_c = \bigcup_{\substack{\vec{w} \in W \\ \vec{p}' \in P}} \{\vec{p} = \mathcal{A}_q(\vec{w}, \vec{p}') \mid q \text{ valid}, \ \text{AD}_{\min}(\vec{p}) < s\} \tag{6.12}$$

The computation of high potential predecessor pairs as defined in Section 4.2.3 is performed element-wise (in fact only scalar multiplications or element-wise add/subtract operations occur):

$$\vec{p}_1 = \left| \frac{\vec{w}_1 2^{r_1 + l_{22}} - (-1)^{s_{12}} (-1)^{s_{22}} \vec{w}_2 2^{r_2 + l_{21}}}{2^{l_{11} + l_{22}} - (-1)^{s_{22}} 2^{l_{12} + l_{21}}} \right| \tag{6.13}$$

$$\vec{p}_2 = \left| \frac{\vec{w}_1 2^{r_1 + l_{12}} - (-1)^{s_{12}} \vec{w}_2 2^{r_2 + l_{11}}}{2^{l_{21} + l_{12}} - (-1)^{s_{22}} 2^{l_{22} + l_{11}}} \right| \tag{6.14}$$

If no single predecessor and no high potential predecessor pair was found, the search is extended to predecessor pairs which are constructed from the MSD representation of $\vec{w}$. Again, this is straight forward as all combinations leading to a reduced AD are evaluated. To give an example, take the vector $\vec{w} = (23, 43)$ which has an MSD representation of $\vec{w} = (10\bar{1}00\bar{1}, 10\bar{1}0\bar{1}0\bar{1})$. It consists of seven non-zeros which results in an adder depth of three according to (6.7). To reduce the AD, each predecessor must have an AD of two (at most 2^2 non-zeros). Hence, all permutations of four non-zeros in $\vec{p}_1$ and $7 - 4 = 3$ non-zeros in $\vec{p}_2$ are evaluated. For example, the pair $\vec{p}_1 = (000000, 10\bar{1}0\bar{1}0\bar{1}) = (0, 43)$ and $\vec{p}_2 = (10\bar{1}00\bar{1}, 0000000) = (23, 0)$ would be a valid pair as well as $\vec{p}_1 = (10\bar{1}000, 10\bar{1}0000) = (24, 48)$ and

$\vec{p}_2 = (00000\overline{1}, 0000\overline{1}0\overline{1}) = (-1, -5)$, which are $\mathrm{odd}(\vec{p}_1) = (3, 6)$ and $\mathrm{odd}(\vec{p}_2) = (1, 5)$ after normalization.

6.2.4 Evaluation of Predecessors

The cost model has to be adjusted to be able to evaluate vector predecessors. For the high-level model, the number of nodes in the adder graph is obtained by counting the elements of the pipeline sets as defined for RPAG (see (4.1) on page 55). The low-level cost model has to be extended to vectors as the coefficient word size is computed differently for vectors. Therefore, the term $\lceil \log_2(w) \rceil$ in (2.26) and (2.27) has to be replaced by the word size which is necessary to represent the sum of all elements of the vector (assuming equal input word sizes), leading to

$$\mathrm{cost}_A(\vec{w}) = B_\mathrm{n} + \left\lceil \log_2 \left(\sum_{n=0}^{|\vec{w}|} w_n \right) \right\rceil , \qquad (6.15)$$

$$\mathrm{cost}_R(\vec{w}) = \frac{1}{K_\mathrm{FF}} \left(B_\mathrm{n} + \left\lceil \log_2 \left(\sum_{n=0}^{|\vec{w}|} w_n \right) \right\rceil \right) . \qquad (6.16)$$

The gain computation is identical to the gain used in RPAG (4.24) (see page 63) but with the generalized cost functions.

6.2.5 Overall Algorithm

The overall algorithm is identical to that of the RPAG algorithm (see Section 4.2.5). The only changes are the generalizations to vectors as described above. In the practical C++ implementation, all basic operators (like +, -, *) and functions (like odd(), $\mathrm{AD_{min}}$(), min(), max(), etc.) were alternatively defined for vectors such that the same implementation works for both problems. Although the PCMM problem is a pure generalization of the PMCM problem, i.e., the PMCM problem is identical to the PCMM problem with vectors of length one, there is of course a performance drop in the implementation when all operations work on vectors with length one in the PMCM case. To avoid this, the implementation uses C++ templates. In the PMCM case, the template type is set to `int` while it is set to `vector<int>` in the PCMM case.

6.3 Experimental Results

To evaluate the performance of the CMM extension, three experiments were made. In the first experiment, the proposed method for the PCMM optimization was compared against using PMCM multiple times, as there is no other CMM algorithm available that respects pipelining. RPAG was used multiple times to compute a PMCM circuit for each column and a pipelined adder tree was used for each output, denoted as "mult. RPAG" in the following. This was performed for a set of 100 instances of random square matrices $N \times N$ each with $N = 2 \dots 10$ and 8 bit coefficients. The results in terms of average number of registered operations (over 100 instances) as well as the improvement of RPAG-CMM over RPAG are shown in Figure 6.3. Significant reductions of registered operations can be achieved which are in a range of $10 \dots 22\%$ except for 8×8 matrices. Here, a better solution is often found by RPAG. This can be explained by the fact that a pipelined adder tree with 2^k inputs can be realized without register overhead. However, even if RPAG performs better for 8×8 matrices on average, in 49 out of 100 cases, a better solution was obtained by the proposed approach nevertheless. Note that this experiment is statistically significant but less realistic as most linear transforms are not random. In fact, they may contain several identical coefficients which appear periodically (e. g., in FFT/DCT).

Therefore, solutions from several publications are taken for comparison in the second experiment. These are analyzed in the number of add/subtract operations, the adder depth and the number of registered operations after careful manual pipelining. For the proposed method, the number of adders is obtained for a CMM_{MAD} optimization, the number of registered operations (add/subtract and pipeline registers) is obtained for a PCMM optimization. The results are listed in Table 6.1, unavailable data is marked with a '–'.

Matrices C_1 and C_2 are color space conversion matrices for RGB to YCbCr and vice versa, taken from the corresponding ITU recommendation [25]. The previous results are taken from Gustafsson and Holm [39] and are reproduced for the sake of comparison. Matrices C_3-C_5 are numeric examples taken from Figure 4 of [150], Figure 2 of [149] and Figure 1 of [37], respectively. Matrices C_6-C_8 are multiplication-free linear transforms as they contain only power-of-two elements of the form $\pm 2^k$ with $k \in \mathbb{N}_0$. Matrix C_6 is used in H.264/AVC video coding and its CMM solution is taken from Figure 1 of [154]. Matrices C_7 and C_8 are used in an 8×8 Walsh-Hadamard transform (also known as the Hadamard transform or Walsh transform) and a (16,11) Reed-Muller error correcting code, respectively. Their CMM solu-

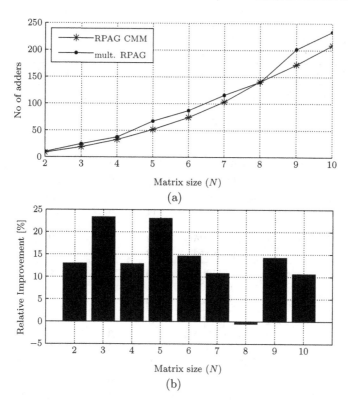

Figure 6.3: Benchmark of random 8 bit $N \times N$ matrices (a) absolute number of adders (b) improvement of RPAG-CMM over multiple use of RPAG

tions are taken from Figures 5 and 6 of [12]. Note that the Walsh-Hadamard matrix provided in the paper (Fig. 5 in [12]) contains errors. The provided solutions were obtained with the correct matrix (as recursively defined as in [12]).

It can be observed that the proposed CMM optimization typically finds a lower or equal number of adders compared to the numerous previous methods although it provides minimal AD which typically increases the number of adders significantly. An average reduction of 12.5% of adders could be achieved compared to the best known solution while providing a significantly lower AD (18.75% less on average). The number of registered operations could be reduced by 38.5% on average, compared to the best known solution. Interestingly, the result for the Walsh-Hadamard transform matrix $\mathbf{C}_7$ requires exactly the same number of additions (24) and registered operations

Table 6.1: Comparison of different CMM methods for benchmark matrices from literature (best marked bold)

Matrix	Size	Method	Adders	AD	Registered Ops
C_1	3×3	[39]	17	5	30
		[149]	18	9	–
		[85]	21	4	–
		[150]	18	6	–
		[151]	18	6	–
		mult. RPAG	20	4	26
		proposed	**12**	**4**	**16**
C_2	3×3	[39, 150]	**11**	5	22
		[149]	14	5	–
		[151]	12	5	–
		mult. RPAG	12	4	19
		proposed	12	**4**	**16**
C_3	4×4	[150]	**14**	7	35
		mult. RPAG	21	4	30
		proposed	**14**	**4**	**19**
C_4	2×2	[149]	**6**	4	12
		mult. RPAG	7	**3**	10
		proposed	**6**	**3**	**7**
C_5	2×2	[37]	**5**	3	8
		mult. RPAG	**5**	**2**	**5**
		proposed	**5**	**2**	**5**
C_6	4×4	[154]	**8**	**2**	**8**
		mult. RPAG	12	**2**	12
		proposed	**8**	**2**	**8**
C_7	8×8	[12]	**24**	–	–
		mult. RPAG	61	4	77
		proposed	**24**	**4**	**24**
C_8	16×16	[12]	43	–	–
		mult. RPAG	56	**3**	56
		proposed	**31**	**3**	**47**

(24) as the fast Walsh-Hadamard transform [23]. Hence, without knowing the iterative construction of the matrix – which was exploited to obtain the fast Walsh-Hadamard transform [23] – the proposed algorithm was able to find a solution with identical complexity.

Table 6.2: Synthesis results for the best reference design of Table 6.1 and the proposed method for CMM min AD. and pipelined CMM

Matrix	Method	CMM min AD			PCMM			
		Slices	f_{max} [MHz]	Dyn. Power [mW]	Slices	f_{max} [MHz]	Slices/ f_{max} [1/MHz]	Dyn. Power [mW]
C_1	ref [39]	74	217.3	27.2	122	**672.5**	0.181	17.4
	proposed	**60**	**256.5**	**18.1**	**80**	619.2	**0.129**	**14.0**
C_2	ref [39]	**61**	**218.4**	20.1	**92**	**702.3**	**0.131**	14.5
	proposed	64	215.5	**17.1**	95	592.8	0.160	15.2
C_3	ref [150]	**67**	167.7	28.6	122	**690.6**	0.177	18.8
	proposed	76	**244.6**	**21.0**	**106**	648.9	**0.163**	**16.1**
C_4	ref [149]	**25**	228.2	11.4	46	626.2	0.073	8.7
	proposed	27	**306.2**	**8.9**	**33**	**651.9**	**0.051**	**7.7**
C_5	ref [37]	**26**	367.0	7.6	41	**701.3**	0.058	8.0
	proposed	27	**494.3**	**7.2**	**30**	651.9	**0.046**	**7.2**
C_6	ref [154]	46	**443.3**	**11.4**	47	**718.9**	0.065	10.4
	proposed	**42**	424.1	11.7	**46**	704.2	0.065	10.4
Avg.	ref	49.8	273.6	17.7	78.3	**685.3**	0.114	13.0
	proposed	**49.3**	**323.5**	**14.0**	**65.0**	644.8	**0.102**	**11.8**
Imp.		1.00%	15.4%	21.0%	17.0%	−6.3%	10.4%	9.2%

In the third experiment, the designs from the second experiment were synthesized. For that, a VHDL code generator was implemented with the help of the FloPoCo library [1]. The input word size was chosen as 12 bit for all designs. The synthesis results for a Xilinx Virtex 6 FPGA (XC6VLX75T-FF784-2) are shown in Table 6.2. All results are obtained after place&route using Xilinx ISE 13.4. All inputs and outputs are registered to enable a fair speed comparison. The power consumption was estimated using Xilinx XPower with 1000 random input values. The dynamic power shown in Table 6.2 consists of the estimated power values for clock, logic and signals. The results show that the proposed method for CMM with minimal AD results in nearly the same slice utilization on average but improves the average performance (f_{max}) and power consumption by 15.4% and 21.0%, respectively. The throughput can be further enhanced by using pipelining. This leads to speedups from 59% up to 312%. This speedup comes along with an average slice increase of 57% for the reference designs compared to the non-pipelined realization whereas it is only 32% for the proposed approach. While the performance is slightly reduced, a slice over f_{max} improvement of 10% is still achieved. As expected from previous work (see Section 2.6.4), less dynamic power is needed for the pipelined designs. Although more slices are used for pipelining, the average dynamic power is reduced by 27% and 34% compared to the non-pipelined reference for the pipelined reference and the proposed PCMM method, respectively.

6.4 Conclusion

An algorithm was presented in this chapter to heuristically solve the PCMM and the CMM_{BAD}/CMM_{MAD} problems. It is based on the RPAG algorithm which was presented in Chapter 4. It is the first time that pipelining and bounded or minimum AD was considered in the CMM optimization. In the PMCM case, the number of registered operations could be reduced by 38.5% on average. Besides this, a lower or equal adder count could be obtained compared to a multitude of previous algorithms with reduced AD leading to average adder and AD reductions of 12.5% and 18.75%, respectively. A synthesis experiment confirmed the expected slice and power reductions for PCMM and minimum AD circuits.

The used vector extensions (most of them adapted from [74]) are general and can be applied to other MCM algorithms that use similar core functions like the $\mathcal{A}$-operation and AD computations to construct CMM solvers. Examples are the H_{cub} algorithm [41] or the DiffAG algorithm [43], two of the best known MCM heuristics. This is left open for future work.

Part II

FPGA Specific MCM Optimizations

7 Combining Adder Graphs with LUT-based Constant Multipliers

All MCM methods discussed so far were based on the reduction of constant multiplications to additions, subtractions and bit shifts. In the following chapters, techniques that realize the constant multiplications using FPGA-specific resources are considered. The inclusion of look-up tables (LUTs) in the PMCM optimization is considered in this chapter.

7.1 Related Work

An efficient way for constant multiplications on FPGAs that uses the look-up tables (LUTs) and fast carry chain was proposed by Chapman [155, 156]. It is often referred to as the KCM method (which originally stands for "Ken Chapman's multiplier" but is sometimes used as "constant (k) coefficient multiplier"). It was later refined by Wirthlin [3] who proposed several FPGA related optimizations like the elimination of redundant LUTs as well as an efficient FPGA mapping for Xilinx Virtex FPGAs where LUT and full adder (FA) use the same slice [3]. An extension of the method to MCM was presented by Faust et al. [157]. They observed that identical LUTs can also be shared between different constant multipliers. It was shown that the maximal number of required LUTs is far less than combinatorially possible. Their benchmark results include a comparison with adder-based MCM, which shows that the LUT-based MCM method requires similar resources but less delay compared to non-pipelined adder graphs of their MinLD MCM algorithm [34] for an input word size of 8 bit on an FPGA with 4-input LUTs.

LUT-based constant multiplication was also used for the multiplication with real [158] and rational [159] constants. As the fixed point representation of rational constants contains periodic bit-patterns identical LUTs may occur that can be shared [159].

By observing the previous work it can be concluded that the complexity of the adder graph MCM method significantly depends on the coefficient value(s) while the complexity of the LUT based approach mainly depends on the input word size. Hence, for a given input word size sometimes one method or the other delivers better results depending on the value of the constant. Therefore, the combination of both methods is analyzed in this chapter by incorporating the LUT-based multipliers in the ILP formulation of the PMCM problem which was described in Chapter 5. The work presented in this chapter was originally published in [139].

7.2 LUT-based Constant Multiplication

The main idea in LUT-based constant multiplications is to split the multiplication into several smaller ones and to add the bit shifted partial results. The small constant multiplications can be constructed to directly fit the input size of the FPGA LUTs yielding to compact designs. Consider a two's complement number x with B_i bits:

$$x = -2^{B_i-1}x_{B_i-1} + \sum_{b=0}^{B_i-2} 2^b x_b \qquad (7.1)$$

If this number is multiplied by a constant c_n with B_c bits, the resulting $B_c \times B_i$ multiplication can be divided into several smaller multiplications of size $B_c \times L$ by rearranging the partial sums:

$$\underbrace{c_n \cdot x}_{B_c \times B_i \text{ Mult.}} = c_n \left(\sum_{b=0}^{B_i-2} 2^b x_b - 2^{B_i-1} x_{B_i-1} \right)$$

$$= c_n \sum_{b=0}^{L-1} 2^b x_b + c_n \sum_{b=L}^{2L-1} 2^b x_b + \ldots$$

$$\ldots + c_n \left(\sum_{b=(K-1)L}^{KL-2} 2^b x_b - 2^{KL-1} x_{KL-1} \right)$$

$$= \underbrace{c_n \sum_{b=0}^{L-1} 2^b x_b}_{B_c \times L \text{ Mult.}} + \underbrace{2^L c_n \sum_{b=0}^{L-1} 2^b x_{b+L}}_{B_c \times L \text{ Mult.}} + \ldots \quad (7.2)$$

$$\underbrace{\ldots + 2^{(K-1)L} c_n \left(\sum_{b=0}^{L-2} 2^b x_{b+(K-1)L} - 2^{L-1} x_{KL-1} \right)}_{B_c \times L \text{ Mult.}}$$

Now, $K = \lceil \frac{B_i}{L} \rceil$ multiplications of size $B_c \times L$ and $K-1$ adders are necessary to compute the constant multiplication. If the input word size B_i is not divisible by L, the input x has to be sign extended to the next larger word size such that $B_i' = KL$ bits. Setting L to the input size of the FPGA LUT allows a direct mapping of the $B_c \times L$ constant multiplier to the FPGA LUTs, each tabulating all partial constant multiplier results. An example of a 4×12 multiplier ($B_c = 4$ bit for the constant and $B_i = 12$ bit for the input) using 4-input LUTs is shown in Figure 7.1.

7.3 LUT Minimization Techniques

Several methods were proposed by Wirthlin to reduce the hardware complexity of the circuit [3]. LUTs can be eliminated in the following cases:

a) Removal of constant LUTs, i.e., LUTs that are always '0' or '1' can be replaced by the corresponding constant.

b) Removal of LUTs which are identical to one of the inputs, i.e., the output can be connected to the corresponding input.

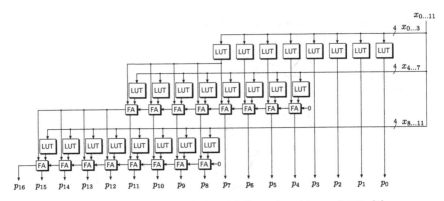

Figure 7.1: 4×12 bit signed multiplier using 4-input LUTs [3]

c) Removal of redundant LUTs, i.e., LUTs with identical content and identical inputs can be shared.

In practice, the synthesis tool will will perform the removal of redundant LUTs when the conditions above occur but the rules are necessary to predict the detailed costs. The removal of redundant LUTs is in particular interesting for the MCM case as it is very likely that redundant LUTs occur [157]. For example, only 52 unique LUTs are sufficient to compute all signed products with constants from 1 to 2^{12} using 4-input LUTs. With increasing LUT input size, this number only increases linear with a factor of about four [157]. This result is rather unexpected as the number of combinations grows exponentially.

7.4 ILP Formulation for the Combined Pipelined Adder/LUT Graph Optimization

PMCM ILP Formulation 2 is extended in such a way that each constant can be alternatively implemented by a LUT-based multiplier while considering the sharing of identical LUTs. For that, we consider a pipelined implementation of the LUT-based multiplier as shown in Figure 7.2. In the first pipeline stage, the LUT content is looked up, shifted and added in the preceding stages using a pipelined adder tree. This structure does not realize the LUT in the same resources used for the RCA like it was proposed in [3]. This is due to the fact that no sharing of LUTs is possible when doing so.

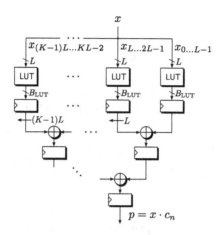

Figure 7.2: Pipelined LUT based multiplier

Note that some of the bit shifts in Figure 7.2 can be moved towards the output of the adder tree to reduce the word size in each stage.

To include LUT-based constant multipliers in the optimization, the set $\mathcal{L}^s_{v,w}$ is defined, which contains all LUT contents of stage s which are needed to compute w from v and a new boolean decision variable is introduced:

$$
l^s_{v,i} = \begin{cases} 1 & \text{if LUT with input node } v \text{ and content } i \text{ is available in stage } s \\ 0 & \text{otherwise} \end{cases}
$$

(7.3)

Now, variables $x^s_{(u,v,w)} \in \{0,1\}$ with $u = 0$ are added to $\mathcal{T}^s$ to model the use of a LUT-based multipler where v and w are the input and output fundamentals, respectively. By changing the objective function and introducing two additional constraints C4 and C5 to PMCM ILP Formulation 2 (the remaining constraints remain nearly identical) the ILP model for the combined optimization is given in ILP Formulation 4.

The two new cost functions cost_{LUT} and $\text{cost}_{\text{AT}}(v,w)$ correspond to the (single output) LUT and the adder tree costs, respectively. This splitting is necessary such that LUTs are counted only once when they are shared between different LUT-based multipliers. The pipeline depth of a LUT-based multiplier, which consists of one stage for the LUT and the depth of the adder tree depends on the input word size of v, denoted $B_v = B_i + \lceil \log_2(v) \rceil$, and the LUT input size L as follows:

$$\text{LD}(v) = \lceil \log_2\left(\lceil B_v/L \rceil\right) \rceil + 1 \tag{7.4}$$

ILP Formulation 4 (Combined PMCM/LUT Mult. Optimization).

$$\min \sum_{s=1}^{S} \sum_{(u,v,w)\in\mathcal{T}^s} \left(\text{cost}(u,v,w) x^s_{(u,v,w)} + \sum_{i\in\mathcal{L}^s_{v,w}} \text{cost}_{\text{LUT}}\, l^s_{v,i} \right)$$

with

$$\text{cost}(u,v,w) = \begin{cases} \text{cost}_{\text{AT}}(v,w) & \text{for } u = 0 \\ \text{cost}_{\text{R}}(w) & \text{for } u = v = w \\ \text{cost}_{\text{A}}(u,v,w) & \text{otherwise} \end{cases}$$

subject to

C1: $\qquad\qquad x^S_w = 1$ for all $w \in T_{\text{odd}}$

C2: $\quad x^s_w - \displaystyle\sum_{(u,v,w')\in\mathcal{T}^s|w'=w} x^s_{(u,v,w)} = 0$ for all $w \in \mathcal{S}^s, s = 2\ldots S$

C3: $\qquad \begin{aligned} x^s_{(u,v,w)} - x^{s-1}_u \le 0 \\ x^s_{(u,v,w)} - x^{s-1}_v \le 0 \end{aligned} \left. \vphantom{\begin{aligned}a\\b\end{aligned}} \right\}$ for all $(u,v,w) \in \mathcal{T}^s$, $s = 2\ldots S$ with $u \ne 0$

C4: $\quad x^s_{(0,v,w)} - x^{s-\text{LD}(v)}_v \le 0 \left.\vphantom{\begin{aligned}a\\b\\c\end{aligned}}\right\}$ for all $(u,v,w) \in \mathcal{T}^s$, $s = 2\ldots S$ with $w \bmod v = 0$, $s \ge \text{LD}(v)$ and $u = 0$

C5: $\quad x^s_{(0,v,w)} - l^s_{v,i} \le 0 \left.\vphantom{\begin{aligned}a\\b\end{aligned}}\right\}$ for all $(u,v,w) \in \mathcal{T}^s$, $i \in \mathcal{L}^s_{v,w}$ with s=2$\ldots$S , $u = 0$

$$x^s_w \in \{0,1\}, \; x^s_{(u,v,w)} \in \mathbb{R}_0, \; l^s_{v,i} \in \{0,1\}$$

The new constraint C4 ensures that fundamental v is available in stage $s - \text{LD}(v)$ if it is needed to compute w in stage s using a LUT-based multiplier. In addition to the input fundamental, the corresponding LUTs has to be available in the same stage which is constrained by C5.

The model above allows that LUT-based multipliers are inserted anywhere between two fundamentals v and w. However, using different input fundamentals v for several LUT-based multipliers is very unlikely to be efficient as no LUTs can be shared. Furthermore, input fundamental $v = 1$ will introduce the least cost for the adder tree as it has the smallest word size. Therefore, the choice of $v = 1$ for LUT-based multipliers will happen most of the time. A similar observation can be made for output fundamentals as if a LUT-based multiplier is used it can directly compute the output fundamental. There are only very rare cases where a LUT-based realization of a NOF may improve the overall cost. This was also observed in experiments. Hence, the search space can be greatly reduced by constraining the inputs and outputs to $v = 1$ and $w \in T_{\text{odd}}$, respectively.

7.5 Experimental Results

The same 2D filter benchmark as in Section 5.6.2 using the same parameters and target FPGA (Virtex 6) is used to evaluate the combined optimization. The Virtex 6 FPGA contains LUTs which can be either configured to single 6-input LUTs or dual 5-input LUTs with shared inputs. As the LUT-based multipliers have shared inputs it is advantageous to chose the input size as $L = 5$. In terms of BLEs, the cost contribution of each 5-input LUT is one half of a BLE each (i. e., $\text{cost}_{\text{LUT}} = \frac{1}{2}$). As single FFs are most of the time mapped to a single BLE the FF costs are estimated with $\text{cost}_{\text{R}}(w) = 1$.

The optimization results in terms of BLEs are listed in Table 7.1 for the optimal PAG method of Chapter 5, the pure LUT MCM method (including pipelining) as well as the presented pipelined adder/LUT graph, denoted as optimal pipelined adder/LUT graph (PALG) method. The coefficients which has to be replaced by LUT multipliers to obtain the PALG solution are given in the last column of Table 7.1. As expected, the LUT MCM method performs better for low input word sizes. Indeed, for $B_{\text{i}} = 8$ it is always the best choice, for $B_{\text{i}} = 10$ it is sometimes the best and for $B_{\text{i}} = 12$ it is never the best method. This is reflected in the combined PALG optimization: For $B_{\text{i}} = 8$, all instances were pure LUT MCM realizations, for $B_{\text{i}} = 10$ there is a mixture of pure adder graphs, pure LUT realizations and combinations of both, while for $B_{\text{i}} = 12$ there are only adder graph realizations. Examples of an optimal PAG and an optimal PALG are given in Figure 7.3. While the PAG in Figure 7.3(a) needs two elements in the second pipeline stage (1 and 7) to compute coefficient 121, this costly coefficient is better realized using a two-stage LUT-based multiplier as shown in Figure 7.3(b). On average, an overall reduction of 5.8% of BLEs compared to the optimal PAG is achieved.

Table 7.1: Optimization results in terms of the number of BLEs for the previous methods RPAG [8] using R iterations and the optimal pipelined adder graphs

Filter Type	Filter Size	B_c	B_i	No. of BLE			
				Optimal PAG	LUT MCM [157]	Optimal PALG	LUT Coeff
gaussian	3×3	8	8	58	**56**	**56**	all
gaussian	5×5	12	8	111	**77**	**77**	all
laplacian	3×3	8	8	61	**54**	**54**	all
unsharp	3×3	8	8	56	**51**	**51**	all
unsharp	3×3	12	8	91	**64**	**64**	all
lowpass	5×5	8	8	98	**91**	**91**	all
lowpass	9×9	10	8	221	**192**	**192**	all
lowpass	15×15	12	8	$\leq$478	**371**	**371**	all
highpass	5×5	8	8	74	**69**	**69**	all
highpass	9×9	10	8	85	**83**	**83**	all
highpass	15×15	12	8	186	**170**	**170**	all
gaussian	3×3	8	10	**68**	71	**68**	none
gaussian	5×5	12	10	129	**98**	**98**	all
laplacian	3×3	8	10	71	69	**68**	all
unsharp	3×3	8	10	**66**	**66**	**66**	none
unsharp	3×3	12	10	105	**80**	**80**	all
lowpass	5×5	8	10	114	118	**112**	33
lowpass	9×9	10	10	255	276	**254**	117
lowpass	15×15	12	10	$\leq$547	546	$\leq$**545**	none
highpass	5×5	8	10	88	95	**87**	121
highpass	9×9	10	10	**101**	115	**101**	none
highpass	15×15	12	10	218	254	**216**	23
gaussian	3×3	8	12	**78**	143	**78**	none
gaussian	5×5	12	12	**147**	203	**147**	none
laplacian	3×3	8	12	**81**	142	**81**	none
unsharp	3×3	8	12	**76**	135	**76**	none
unsharp	3×3	12	12	**119**	173	**119**	none
lowpass	5×5	8	12	**130**	242	**130**	none
lowpass	9×9	10	12	**289**	589	**289**	none
lowpass	15×15	12	12	$\leq$**620**	1203	$\leq$**620**	none
highpass	5×5	8	12	**102**	192	**102**	none
highpass	9×9	10	12	**117**	232	**117**	none
highpass	15×15	12	12	**250**	523	**250**	none
			Average:	160.3	207.36	150.97	
	Improvement to opt. PAG:			–	-29.36%	5.82%	

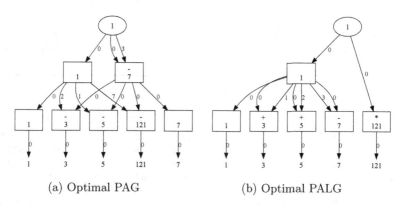

(a) Optimal PAG   (b) Optimal PALG

Figure 7.3: Example PAG/PALG of the highpass 5×5 instance with $B_i = 10\,\text{bit}$

Like for the optimal PAG method, an optimal solution could be found except for two of the lowpass 15×15 instances. The computation time for solving the extended PALG formulation was comparable to that of the PAG formulation and even less (2129.79 seconds on average for PALG compared to 3538.58 seconds for PAG).

Synthesis experiments were performed for the same FPGA and tools as used in Section 5.6.2. For that, the VHDL code generator was extended to PALGs. The results including the optimal PAG results and pure LUT-based MCM results are listed in Table 7.2. The synthesis results confirm that the PALG method leads to the least resources in most of the cases. On average, the BLEs are reduced by 7.38% compared to the optimal PAG by including LUT-based multipliers in the optimization. However, like for the PAG method, the cost model leads to more resources than needed in some of the cases. Due to the larger bit shifts in the LUT-based multiplier (see Figure 7.2), the LSBs in the adder tree are pure flip-flops. Hence, it is likely that the costs for a LUT-based multiplier are overestimated.

7.6 Conclusion

An ILP formulation for the PMCM problem which includes LUT-based constant multipliers in the PAG was presented in this section. It was shown that this combined optimization can further reduce the resources. From the obtained results, a combined optimization seems to be useful for a limited

Table 7.2: Synthesis results using the same benchmark instances as in Table 7.1 providing the actual BLEs as well as the maximum clock frequency f_{max} (best marked bold) on a Virtex 6 FPGA

Filter Type	Filter Size	B_c	B_i	Optimal PAG BLE	f_{max}	LUT MCM [157] BLE	f_{max}	Optimal PALG BLE	f_{max}
gaussian	3 × 3	8	8	**58**	730.5	**58**	720.5	**58**	720.5
gaussian	5 × 5	12	8	111	619.2	**68**	633.3	**68**	633.3
laplacian	3 × 3	8	8	61	701.8	**52**	718.9	**52**	718.9
unsharp	3 × 3	8	8	56	709.7	**51**	731.0	**51**	731.0
unsharp	3 × 3	12	8	91	657.9	**59**	701.3	**59**	701.3
lowpass	5 × 5	8	8	98	657.9	**93**	652.3	**93**	652.3
lowpass	9 × 9	10	8	221	582.4	**186**	573.1	**186**	573.1
lowpass	15 × 15	12	8	478	489.0	**368**	525.8	**368**	525.8
highpass	5 × 5	8	8	74	711.7	**67**	726.2	**67**	726.2
highpass	9 × 9	10	8	85	689.2	**81**	655.3	**81**	655.3
highpass	15 × 15	12	8	183	598.4	**166**	630.5	**166**	630.5
gaussian	3 × 3	8	10	68	745.7	70	776.4	**64**	666.7
gaussian	5 × 5	12	10	129	637.8	**82**	690.6	**82**	690.6
laplacian	3 × 3	8	10	71	656.6	**59**	745.2	**59**	745.2
unsharp	3 × 3	8	10	66	712.3	**60**	769.8	66	712.3
unsharp	3 × 3	12	10	103	605.3	**69**	690.1	**69**	690.1
lowpass	5 × 5	8	10	114	646.8	**106**	627.0	111	647.3
lowpass	9 × 9	10	10	251	602.1	**225**	574.4	252	528.3
lowpass	15 × 15	12	10	546	514.7	**443**	414.8	529	470.8
highpass	5 × 5	8	10	86	704.2	89	698.8	**87**	721.5
highpass	9 × 9	10	10	97	677.5	**91**	644.8	97	677.5
highpass	15 × 15	12	10	218	567.9	218	600.6	**203**	583.8
gaussian	3 × 3	8	12	**78**	644.3	129	688.2	**78**	644.3
gaussian	5 × 5	12	12	**147**	624.2	150	615.4	**147**	635.7
laplacian	3 × 3	8	12	**81**	724.1	104	672.0	**81**	724.1
unsharp	3 × 3	8	12	**72**	638.6	129	638.2	**72**	638.6
unsharp	3 × 3	12	12	**119**	638.6	122	620.0	**119**	638.6
lowpass	5 × 5	8	12	**126**	653.2	198	557.1	130	625.8
lowpass	9 × 9	10	12	278	581.7	412	606.4	**273**	602.4
lowpass	15 × 15	12	12	**619**	530.8	835	530.8	620	517.9
highpass	5 × 5	8	12	**94**	714.3	145	647.7	**94**	714.3
highpass	9 × 9	10	12	**113**	680.7	159	624.6	**113**	680.7
highpass	15 × 15	12	12	**239**	539.7	388	593.5	250	590.7
			avg.:	158.52	642.08	167.64	645.28	146.82	648.94
	Improvement to Opt. PAG:			–	–	−5.75%	0.49%	7.38%	1.07%

range of the input word size; for a smaller input word size, the LUT-based multipliers performed better while for larger word sizes the PAGs are the better choice. However, the benchmark was limited to small coefficient word sizes due to the computational complexity of the exact method. Indeed, the best results for the LUT-based constant multipliers were achieved for the largest coefficient word size of 12 bit. Hence, the consideration of LUT-based constant multipliers in a heuristic like the RPAG algorithm that is able to deal with large coefficient word sizes would be interesting but is left open for future work.

8 Optimization of Hybrid Adder Graphs Containing Embedded Multipliers

The demand of multiplication intensive applications like those in the domain of digital signal processing was the driving force for embedding multipliers as hard blocks into the fabric of FPGAs. Modern FPGAs also include pre and post adders around the multiplier which is then commonly called a DSP block. These may be, however, of limited use due to the following reasons:

- Limited quantity: The number of DSP blocks is limited and may either not be sufficient or occupied by other parts of the system.

- Fixed word size: If a larger word size is required, several DSP blocks may be necessary. If a smaller word size is required, large parts of the DSP block may be unused.

- Power consumption: As there is no sharing of intermediate results like in adder graph based MCM, all partial products have to be computed for each coefficient which increases the switching activity within the DSP blocks.

It was shown in the previous chapters that MCM operations can be mapped well to FPGAs by using the logic and carry-chain resources. Therefore, a method that is able to include a user-specified number of embedded multipliers into the PMCM optimization which is included in the RPAG algorithm is proposed in this chapter. This extension is available in the open-source RPAG implementation [16]. It can be used to reduce the logic resources by realizing the most complex coefficients using DSP blocks or to reduce the number of DSP blocks due to sharing of intermediate results in large word size applications. These large word sizes typically appear in applications in which floating point arithmetic is used. Even if it is much more resource intensive, more and more companies use floating point on FPGAs due to the reduced time-to-market (no fixed-point word size analysis necessary) and a simplified verification process.

The main complexity in floating point multiplications with constants is found in the integer multipler block for the mantissa, for which the extended RPAG algorithm is used for complexity reduction. This is presented in the following sections of this chapter and was originally published in [160]. An architecture for an MCM operation in floating point arithmetic will be addressed in Chapter 9.

8.1 Implementation Techniques for Large Multipliers

The mantissa multiplications with constants correspond to an integer MCM operation which should be pipelined for performance reasons. Thus, it can be realized using the PMCM techniques presented so far. The main difference compared to fixed point coefficients is the increased word size which increases the complexity as well as the pipeline depth. The floating point standard [161] requires a mantissa word size for single and double precision of 24 bit and 53 bit (including leading one), respectively. In the worst-case MSD representation every second bit is a non-zero except the two MSBs which are both non-zero (see (4.26) on page 69). This leads to a maximum adder depth of four and five for single and double precision, respectively. While this is problematic for the optimal method, it is still feasible for the RPAG heuristic.

However, using the embedded multipliers or DSP blocks of FPGAs is not that straight forward due to their limited word size. Take for example, the 17×24 bit multipliers (unsigned) of the DSP48(E) blocks of Xilinx' Virtex 5/6/7 FPGAs. A 24×24 bit multiplication as necessary for single precision requires to cascade two DSP blocks as shown in Figure 8.1(a). Argument a is split into two smaller words fitting the input word size of 17 bit and the results are bit-shifted and added by using the post-adder of the DSP block.

In case that one argument is a constant, there are much more possibilities to distribute the constant into lower words by allowing different shifts. The constant can be represented as

$$c = c_{\text{Low}} \pm 2^B c_{\text{High}} \qquad (8.1)$$

leading to the structure shown in Figure 8.1(b). In principle, there are many combinations of c_{Low} and c_{High} as long as the constant is less than the total

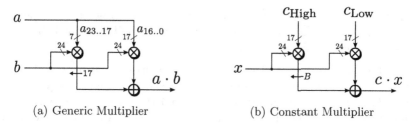

(a) Generic Multiplier (b) Constant Multiplier

Figure 8.1: A single precision mantissa multiplier using DSP48E blocks on a Xilinx Virtex 6 FPGA

input word size (which is $2 \cdot 17$ bit $= 34$ bit in this example). To illustrate this, consider the 24 bit constant 11508221 which can be split into the following $c_{\text{Low}}/c_{\text{High}}$ combinations:

$$11508221 = 104957 + 2^{17} \cdot 87 \tag{8.2}$$
$$= 39421 + 2^{16} \cdot 175 \tag{8.3}$$
$$= 6653 + 2^{15} \cdot 351 \tag{8.4}$$
$$= 39421 + 2^{15} \cdot 350 \tag{8.5}$$
$$= 72189 + 2^{15} \cdot 349 \tag{8.6}$$

$$\vdots$$

This degree of freedom can be used to find $c_{\text{Low}}/c_{\text{High}}$ combinations which can be shared among different output coefficients.

Going to larger word sizes, e.g., as required for double precision, requires even more DSP blocks. A classical method to reduce the required number of multiplications at the cost of additional additions can be obtained by using the Karatsuba-Ofman algorithm [162]. However, it was shown by de Dinechin and Pasca that a direct application of the Karatsuba-Ofman algorithm may lead to poor mappings on modern FPGAs due to limitations of the DSP blocks [163]. They proposed a method called *tiling* instead, in which an $n \times k$ multiplier is graphically represented as an $n \times k$ rectangle. Then, smaller multipliers, which are represented by corresponding smaller rectangles, are placed into the larger one until it is completely covered. All multiplier results have to be bit-shifted and added. The bit shifts can be directly computed by evaluating the position of the corners of the multiplier rectangles. The goal is then to find the most compact configuration. First manual optimizations with application to a 53×53 bit mantissa multiplier were presented in [163], an automated method was proposed in [4]. The post-adders of DSP blocks can also be incorporated to build so-called super-tiles which may have a different shape [4].

A tiling of a double precision 53×53 bit mantissa multiplier is shown in Figure 8.2(a) [4], the corresponding circuit is shown in Figure 8.2(b). It uses seven 17×24 DSP multipliers (M1-M7) and three small multipliers using logic, two 17×6 (M8-M9) and one 20×2 (M10). All adders except the two rightmost adders are post-adders from the DSP blocks. For comparison, the intellectual property (IP) core from Xilinx requires ten DSP blocks plus additional logic for the same operation. Again, the situation is different in case that one argument is a constant. Three 17×24 DSP multipliers can be cascaded with a small 2×24 logic multiplier to get a 53×24 multiplier. Now, partial products with 24 bit constants can be computed by a layer of DSP blocks. These values can be shared among different outputs which are computed using additional adders.

A similar conclusion can be drawn in cascading the DSP blocks of FPGAs from Altera. Their latest FPGAs (the Stratix V, Cyclone V and Aria V devices) provide so-called variable precision DSP blocks. Each of them can be configured as three independent 9×9 bit multipliers, two independent 16×16 bit, 15×17 bit or 14×18 bit multipliers, or a single 18×36 bit or 27×27 bit multiplier.

From the observations above, an optimization problem can be derived to find the minimum set of partial products with limited word size, such that a given set of target constants can be computed by combining those using a minimum adder graph. It may be noted that this problem is very similar to the steps performed during the RPAG optimization. While RPAG is forced to reduce the AD in each stage, the word size has now to be reduced to the multiplier input size when using the DSP blocks. Therefore, the RPAG algorithm is extended in the next section to include a user-defined number of embedded multipliers in the pipelined adder graph (PAG).

8.2 Hybrid PMCM Optimization with Embedded Multipliers

This section addresses the inclusion of embedded multipliers into the PAG which is called hybrid PAG in the following. It can be used to reduce the logic resources of a PAG by inserting as many embedded multipliers or DSP

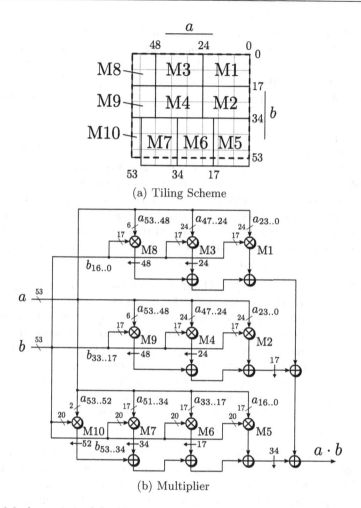

(a) Tiling Scheme

(b) Multiplier

Figure 8.2: An optimized double precision mantissa multiplier on a Xilinx Virtex 6 FPGA [4]

blocks as available for the FPGA or application. DSP block results are efficiently shared, such that the method is also useful to reduce the amount of DSP blocks in MCM operations. An example of a hybrid PAG is given in Figure 8.3. It is separated into three parts: 1) the input PAG, 2) the embedded multipliers, and 3) the output PAG. The input PAG has the same pipeline depth as the embedded multipliers, denoted as S_i. Differences

in the pipeline depths of embedded multipliers and the input PAG can be
simply balanced by placing additional registers at the input. The output
PAG has S_o pipeline stages, leading to a total of $S = S_i + S_o$ pipeline
stages. The input PAG can compensate for (possibly) insufficient embedded
multipliers and the output PAG is used to compute output coefficients with
a word size larger than the multiplier word size.

The required optimization for hybrid PAGs can now be decomposed into
three parts: First, the output PAG is optimized where the objective is to
reduce the word size of the target set to the word size of the embedded
multipliers. Second, the fundamentals with reduced word size have to be
assigned to embedded multipliers or the input PAG. And third, the optimization of the input PAG, where the objective is to reduce the adder depth
to zero. Of course, parts may be skipped if not needed, e. g., for low word
size coefficients there is no output PAG, for a large number of multipliers
there may be no input PAG necessary.

During the search in the RPAG optimization (see Section 4.2.1), predecessors are evaluated that reduce the adder depth of the working set elements.
Reducing the adder depth does not necessarily reduce the word size. Take
for example, a binary number of the form $100\ldots0001$. It has an adder depth
of one but can have an arbitrary large word size. Thus, an additional word
size (B_s) constraint is introduced for the predecessor selection. Both, word
size and existing adder depth constraints can be set to arbitrary values depending on the stage of the hybrid PAG. Constraining the word size always
limits the AD. Hence, the word size constraint is more rigorous in limiting
the search space than the AD constraint.

The constraints are different for different segments of the hybrid PAG.
Four different segments can be separated which are shown as bars at the right
side of Figure 8.3. The corresponding constraints are listed in Table 8.1. In
the first segment, which corresponds to all stages of the input PAG except
the last stage, only the adder depth is constrained to the current stage s.
In the following segments II-IV, only the word size has to be constrained.
Clearly, in segment II (stage S_i) it has to be constrained to the input word
size of the embedded multipliers B_M where the constant is applied. With
each following stage, the word size can be at most doubled, leading to the
upper bound $B_s^{\max}$ in Table 8.1. The output word size (segment IV) is
obviously at most the target word size B_T. With each previous stage, the
word size can be halved in the best case, leading to the lower bound $B_s^{\min}$.
Note that there is a degree of freedom in choosing B_s as long as we have the
common case that B_T is not divisible by B_M. A lower B_s tends to reduce

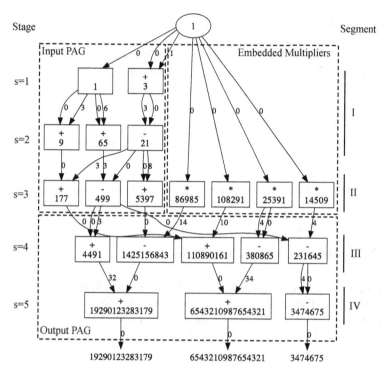

Figure 8.3: A pipelined hybrid adder graph consisting of input adder graph, embedded multipliers $(S_i = 3)$ and output adder graph $(S_o = 2)$

Table 8.1: Constraints for adder depth AD_s, minimum word size $B_s^{\min}$, and maximum word size $B_s^{\max}$ of the segments shown in Figure 8.3

Segment	Stage s	AD_s	$B_s^{\min}$	$B_s^{\max}$
I	$0 \ldots S_i - 1$	s	0	∞
II	S_i	∞	B_M	B_M
III	$S_i + 1 \ldots S_o - 1$	∞	$\left\lceil \frac{B_T}{2^{S-s}} \right\rceil$	$2^{s-S_i} B_M$
IV	$S = S_i + S_o$	∞	B_T	B_T

the number of elements at lower stages, and vice versa. Therefore, a lower number of embedded multipliers can be achieved at the cost of additional adders by choosing a lower B_s (and vice versa).

In the following, the minimal necessary pipeline depth for the input and output PAGs are derived.

8.2.1 Depth of the Input PAG

The depth of the input PAG depends on the maximum number of non-zero digits of the constants. For a B bit binary number, the resulting adder depth is limited by (4.26) (see page 69), leading to a pipeline depth of the input PAG of

$$S_i = \log_2(\lfloor B/2 \rfloor + 1) \ . \tag{8.7}$$

Hence, a multiplier with word size B can be replaced by a PAG with S_i stages.

8.2.2 Depth of the Output PAG

For target coefficients for which $B_T > B_M$, an output PAG is required. The input and output word sizes of the output PAG are B_M and B_T, respectively. As the word size can be doubled by each stage of adders the sufficient number of stages needed for the output PAG is given by

$$S_o = \lceil \log_2 (B_T / B_M) \rceil \ . \tag{8.8}$$

8.3 RPAG Modifications

The RPAG algorithm was extended by the word size and adder depth constraints as defined in the previous sections. By specifying the maximum number of embedded multipliers that can be used in the hybrid PMCM solution (`--max_no_of_mult`) and the multiplier word size (`--mult_wordsize`), the depth of input and output PAG are obtained and the constraints are set accordingly. Per default, the word size constraint B_s is set to the average of $B_s^{\min}$ and $B_s^{\max}$ if not defined different by the user. After the optimization of the output PAG (in stage S_i), the assignment to the embedded multipliers have to be performed. For that, the elements of X_{S_i} with the highest non-zero count are selected as these have the highest probability to produce the highest cost in the input PAG. The selected elements are assigned to embedded multipliers and are removed from X_{S_i}. This may reduce the pipeline depth of the input PAG which is evaluated after the reduction of X_{S_i}.

8.4 Experimental Results

The hybrid optimization is evaluated by synthesis experiments of different use cases. For that, the Matlab-based VHDL code generator was extended to include embedded multipliers. All synthesis results were obtained using the Xilinx ISE 13.4 for the Virtex 6 FPGA XC6VSX475T-1 (after place & route). All obtained PMCM blocks were designed with full-precision, i.e., no truncation was performed. Dynamic power estimations were obtained using using XPower with simulation results for an input sequence of 1000 uniformly distributed random samples.

8.4.1 Trade-off Between Slices and DSP Resources

The trade-off between slices and DSP blocks is evaluated in first experiment by using the low word size benchmark coefficients of Mirzaei [5], which were already used in previous Chapters. For that, the number of embedded multipliers M is equally varied between zero and the number of unique coefficients N_{uq} in 5 steps. Table 8.2 lists the resulting number of registered operations as well as the slice resources. A fine-grained plot of the slices and the dynamic power dissipation for filter instance $N = 28$ and $M = 0, 1, 2, \ldots, N_{uq}$ is shown in Figure 8.4. It can be observed, that a nearly linear reduction of slice resources can be achieved by using more embedded multipliers. The dynamic power consumption follows the opposite trend. For this example, 61% more dynamic power is required by using only embedded multipliers compared to the slice implementation. As far as the power estimation models of XPower are accurate enough, it seems that the higher switching activity in embedded multipliers dominates the increased routing power of the slice-based PMCM circuit.

8.4.2 Reducing Embedded Multipliers in Large MCM Blocks

In the second experiment, the hybrid PAG optimization was used to reduce DSP block resources for large word size single and double floating-point mantissa MCM blocks. For that, a set of seven FIR filters was designed with 10 to 120 taps. All filters were designed as low pass filters using the Parks-McClellan algorithm with a passband and stopband frequency of $f_p = 0.3$ and $f_s = 0.4$ (normalized to Nyquist frequency), respectively. The resulting

Table 8.2: Synthesis results of benchmark set [5]

$M =$		0		$\lfloor \frac{1}{4} N_{\mathrm{uq}} \rfloor$		$\lfloor \frac{1}{2} N_{\mathrm{uq}} \rfloor$		$\lfloor \frac{3}{4} N_{\mathrm{uq}} \rfloor$		N_{uq}	
N	N_{uq}	reg. ops	Slices	reg. ops	Slices	reg. ops	Slices	reg. ops	Slices	reg. ops	Slices
6	2	10	42	10	42	7	27	7	27	3	13
10	4	13	60	10	47	7	25	5	20	3	10
13	6	18	84	14	65	9	41	6	26	3	13
20	8	19	95	13	60	10	46	5	19	3	11
28	13	23	114	20	96	16	80	9	37	3	16
41	20	32	166	25	125	16	84	9	47	3	10
61	30	53	218	34	174	24	120	15	73	3	14
119	53	70	334	52	241	34	153	16	61	3	13
151	70	86	401	69	320	43	204	24	113	3	14
avg.:		36.0	168.2	27.4	130.0	18.4	86.7	10.7	47.0	3	12.7

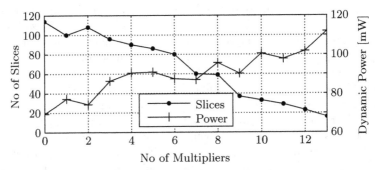

Figure 8.4: Pareto front for filter $N = 28$ of benchmark set [5]

passband/stopband approximation errors ranges from $-15\,\mathrm{dB}$ (10 taps) to $-100\,\mathrm{dB}$ (120 taps). The coefficient values of the integer mantissas are given in Appendix A.2. Due to the symmetry of the coefficients and the large word size, the number of unique coefficients N_{uq} is exactly half the number of taps. The optimized hybrid PMCM blocks were synthesized and compared with state-of-the-art methods. The synthesis and dynamic power results for the corresponding single and double precision MCM blocks are listed in Table 8.3 and Table 8.4, respectively.

For single precision, each constant multiplication was performed using two 17×24 DSP blocks as reference (see Figure 8.1(a)). Here, both methods use the same number of adders as one adder per output coefficient is required to merge DSP block results. Comparing the DSP count, it can be observed that a large number of DSP blocks can be saved in the optimized hybrid PAG due to sharing. For large N_{uq}, only about half of the DSP blocks are necessary. The average dynamic power could be reduced by 8.4%. However, the adders of the conventional method can be completely mapped to the post-adders of the DSP blocks. The DSP48E(1) slice of the Virtex 5/6/7 devices contains a 48 bit post-adder including a hard wired 17 bit input shift. The post-adders can only be used in hybrid PAGs when the result of a multiplier is not used by other coefficients. This is due to a routing limitation of the DSP48E(1) slice whose single output port can be set to either the multiplier result *or* the post-adder result. Hence, halving the DSP count comes at the expense of additional slices for post-addition. This is illustrated in Figure 8.5 which shows the conventional method and the resulting hybrid RPAG solution for the smallest filter instance, both drawn as a hybrid PAG. Only two of the adders in Figure 8.5(b) (computing factors 4552139 and 6910239) can be realized using the post adders while all adders shown in Figure 8.5(a) can be mapped to post adders.

For double precision, the optimized multiplier of Banescu et al. [4] as shown in Figure 8.2(b) was used for each coefficient as reference. It can be generated by FloPoCo [2]. Interestingly, one out of the seven DSP block multipliers was realized using slice logic in the generated core. For the hybrid PAG optimization, a 53×24 multiplier using three DSP blocks (each used for a 17×24 multiplication and post addition) and some additional logic (a 2×24 multiplier, realized using AND-gates and one addition) was used. Therefore, the extended RPAG algorithm was configured to reduce the constants from 53 bits at the output stage to 24 bit within two stages. The results are listed in Table 8.4. On average, slice, DSP and dynamic power reductions of 40%, 15% and 35% could be achieved while a similar speed was obtained. It can be observed that larger coefficient counts typically lead to higher reductions in terms of slice, DSP and power.

8.5 Conclusion

A method for incorporating a user-specified number of embedded multipliers in the heuristic PMCM optimization of the RPAG algorithm was introduced. With that, a trade-off between DSP slices and logic can be obtained. Due to the efficient sharing of embedded multipliers, the number of DSP blocks can be reduced in MCM instances with large word size coefficients that typically appear in floating point applications. For the considered single precision benchmark, the DSP reductions lead to power and delay reductions which

Table 8.3: MCM benchmark with single precision mantissas

N_{uq}	MCM using conventional tiling					Hybrid RPAG (proposed)				
	add ops	DSP	Slices	Pow. [mW]	f_{max} [MHz]	add ops	DSP	Slices	Pow. [mW]	f_{max} [MHz]
5	5	10	25	53	252	5	7	62	55	332
10	10	20	50	89	239	10	17	117	97	308
20	20	40	100	154	240	20	23	238	150	297
30	30	60	150	221	224	30	30	353	204	281
40	40	80	200	293	220	40	40	451	260	277
50	50	100	253	363	212	50	50	578	323	252
60	60	120	313	436	212	60	59	696	384	334
avg.:	30.7	61.4	155.9	229.8	229.0	30.7	32.3	356.4	210.4	297.7
imp.:	–	–	–	–	–	0%	47.4%	−128.6%	8.4%	30%

Table 8.4: MCM benchmark with double precision mantissas

N_{uq}	MCM using tiling of [4]					Hybrid RPAG (proposed)				
	add ops	DSP	Slices	Pow. [mW]	f_{max} [MHz]	add ops	DSP	Slices	Pow. [mW]	f_{max} [MHz]
5	25	30	951	430	203	15	30	642	286	221
10	50	60	1704	789	196	30	60	1233	538	227
20	100	120	3411	1659	192	60	111	2441	1013	195
30	150	180	5784	2539	178	90	171	3707	1616	173
40	200	240	7483	3471	175	120	204	4489	2251	172
50	250	300	9411	4181	177	150	243	5684	2948	135
60	300	360	10403	5200	149	180	273	6367	3148	169
avg.:	153.6	184.3	5592.4	2610.0	181.8	92.1	156.0	3509.0	1686.0	185.3
imp.:	–	–	–	–	–	40.0%	15.4%	37.3%	35.4%	1.9%

comes along with an increase of slices. For the double precision benchmark, all performance metrics could be improved. Note that the proposed hybrid PMCM optimization originally presented in [160] was recently modified to the non-pipelined hybrid MCM problem by Aksoy et al. [164].

Further investigations could be performed towards a better usage of the post adders of DSP slices. These can not be used if the multiplier output is shared among different outputs. Extending the DSP48E block by a separate output or an additional cascading path for distributing multiplier results between different DSP blocks would be a great extension for large MCM applications in future FPGA generations. To circumvent this problem in current FPGAs, one solution in the single precision case could be in realizing

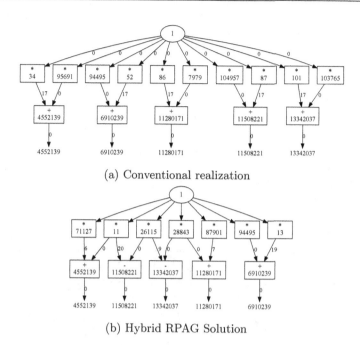

(a) Conventional realization

(b) Hybrid RPAG Solution

Figure 8.5: Hybrid adder/multiplier graph adder graph for the $N_{uq} = 5$ benchmark filter

the least complex part of the constant of (8.1) in slice logic. This can be realized by forcing the hybrid RPAG algorithm to use only one embedded multiplier for each of the N_{uq} unique coefficients. This is illustrated for the $N_{uq} = 5$ example in Figure 8.6. All output adders can use the post addition of the DSP slice, thus, only five adders and two registers with relatively small word size are required in addition. In contrast to that, the scheme in Figure 8.5(b) requires three large word size adders in slice logic to realize the 24 bit factors. This requires a significant higher BLE count (about 600 compared to 150, assuming an input word size of 24 bit) and two additional DSP slices. It can be concluded that besides the large DSP block reductions obtained, there is still room for improvement when vendor specific FPGA DSP block capabilities are exploited in the optimization.

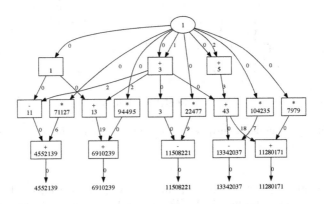

Figure 8.6: Adder graph for the $N_{uq} = 5$ benchmark filter with maximum post adder usage

9 Floating Point Multiple Constant Multiplication

It was shown in the previous chapter that the hardware complexity of the mantissa multiplier block can be reduced by sharing intermediate results. This chapter deals with the remaining architecture to perform a floating point MCM operation, namely the exponent computation, normalization, rounding and the treatment of special values. This architecture with parts of the last chapter was originally published in [165].

9.1 Constant Multiplication using Floating Point Arithmetic

The IEEE floating point representation [161] of a number x is encoded using a sign bit s_x, a normalized mantissa f_x (without the leading one) and an exponent e_x, representing the value

$$x = (-1)^{s_x} \cdot (1.f_x) \cdot 2^{e_x - E_{\text{bias}}} \quad . \tag{9.1}$$

A floating point multiplication $p_n = x \cdot c_n$ can be performed by multiplying the mantissas, adding the exponents, subtracting a bias (E_{bias}) and computing the XOR of the sign bits:

$$1.f'_{p_n} = 1.f_x \cdot 1.f_{c_n} \tag{9.2}$$

$$e'_{p_n} = e_x + e_{c_n} - E_{\text{bias}} \tag{9.3}$$

$$s_{p_n} = s_x \oplus s_{c_n} \tag{9.4}$$

The mantissa f'_{p_n} and exponent e'_{p_n} may be modified by rounding and normalization to get the final results f_{p_n} and e_{p_n}. As the values of both input mantissas $1.f_x$ and $1.f_{c_n}$ lie in the interval $[1, 2)$, the product mantissa lies in the range $1.f'_{p_n} \in [1, 4)$. Hence, if normalization is necessary, it can be done by right shifting the mantissa and incrementing the exponent. Besides these computations, additional logic will be necessary to handle special values like 0, $\pm\infty$ or not-a-number (NaN).

If c_n is a single constant, (9.2) becomes an SCM operation, (9.3) becomes a constant addition with $e_{c_n} - E_{\text{bias}}$ and (9.4) can be realized by either a simple wire (in case $s_{c_n} = 0$) or an inverter (in case $s_{c_n} = 1$). The floating point SCM (FPSCM) operation was examined in detail by Brisebarre et al. [48]. If there are multiple constants of c_n, denoted floating point MCM (FPMCM) in the following, resources can be shared to reduce the hardware complexity:

- Redundancies between several constant integer mantissas can be utilized (the SCM is extended to an MCM).

- Equal constant exponent computations can be shared.

- Special values of the input have to be detected only once.

9.2 Floating Point MCM Architecture

An architecture that considers the observations above is shown in Figure 9.1 for the example of the 32 bit single precision format [161]. First, the input x is split into sign, exponent and mantissa which are processed in parallel. The hidden bit is appended to the mantissa and the mantissa MCM computation (9.2) is done in a pipelined integer multiplier block, producing N results. Each product is rounded, normalized and combined with the corresponding exponent and sign in a post-processing stage. Special values are detected at input side and forwarded to the post-processing stages. The exponent computation, post-processing and integer multiplication are described in the following.

9.2.1 Multiplier Block

The mantissa multiplications with constants correspond to an integer MCM operation for which the PMCM techniques presented in previous chapters can be used.

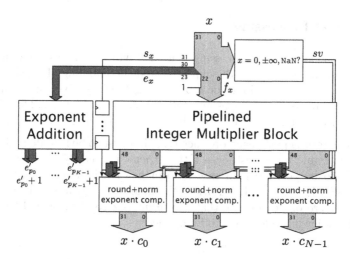

Figure 9.1: Proposed architecture for the single precision floating point MCM operation

9.2.2 Exponent Computation

In principle, there are two possibilities for computing the exponent. The first (and obvious) way is to use N adders to add the constant offset consisting of exponent and bias $(e_{c_n} - E_{\text{bias}})$ plus an optional increment in case of a normalization right shift. The optional increment can be realized using the carry-in of the same adder. The second way is to compute both alternatives for all constant exponents, i. e., $e_{c_n} - E_{\text{bias}}$ and $e_{c_n} - E_{\text{bias}} + 1$, and to select the right one using a MUX. As shown in the next subsection, unused input ports of the exponent multiplexers in the post-processing can be used for this selection without extra resources. Now, the second way becomes more efficient if there are many equal exponents or exponents that differ by only one. More precisely, exponent adders can be saved by using the second way if the condition

$$\left| \bigcup_{n=0}^{N-1} \{e_{c_n}, e_{c_n} + 1\} \right| < N \tag{9.5}$$

is met. For real-live FIR filters, this condition is often met as the filter coefficients have a similar order of magnitude.

9.2.3 Post-Processing

For each output, a post-processing is needed which includes rounding, normalization, the selection of the exponent and the handling of special values. The architecture of a single post-processing block is given in Figure 9.2. There are three pipeline stages in total (not shown in Figure 9.2 for clarity), one after rounding, one after normalization and one after the multiplexer.

In the mantissa path, each product $1.f'_{p_n}$ has first to be rounded. Here, the standard "round to nearest, even if tie" scheme is used [166]. Then, a right-shift has to be performed in case a normalization is necessary (depending on the MSB bit $norm$). In addition, the final mantissa has to be set to bit vectors of zero or one in case that the result is zero, $\pm\infty$ or NaN [161]. For that, the following rules have to be applied:

$$0 \cdot c_n = 0 \tag{9.6}$$

$$\pm\infty \cdot c_n = \pm\infty \cdot (-1)^{s_{c_n}} \tag{9.7}$$

$$\text{NaN} \cdot c_n = \text{NaN} \tag{9.8}$$

Overflows and underflows of the exponent must result in $(-1)^{s_x} \cdot \infty$ and zero, respectively. The detection is done in the block 'exception handling', which computes the control signals of the exponent and mantissa MUXs depending on special values of the input sv and overflows/underflows of the exponent. A 4:1 MUX is used to select between these four cases as it can be directly mapped to a single six-input LUT of modern FPGAs.

The exponent path follows the exponent sharing approach of the last subsection. A MUX is necessary to select between special values ('11...11' vector in case of NaN or $\pm\infty$, 0 in case of a zero output value) and the computed exponent. As mentioned before, a six-input LUT is able to implement a 4:1 MUX, so the same MUX can be used to select the right exponent. Denormalized values are not supported in this architecture as they require additional hardware and are of little use in DSP applications.

9.3 Experimental Results

A VHDL code generator was implemented using Matlab that generates the architecture as presented in figures 9.1 and 9.2 including the PMCM and hybrid PMCM optimization of chapters 4 and 8 and the proposed exponent sharing (when criteria (9.5) is met). The architecture is IEEE 754 compliant [161] except for the support for denormalized values.

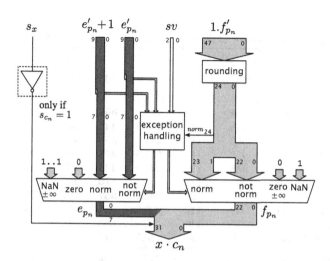

Figure 9.2: Detailed architecture of one round, normalize and exponent selection block of Figure 9.1

9.3.1 Floating Point SCM

In the first experiment, the generated cores are compared to the floating point SCM method of Brisebarre [48], to get a comparison to the state-of-the-art. It is part of the FloPoCo project [2] which also provides a VHDL code generator. Internally, FloPoCo uses a non-standard floating point representation which uses two extra bits for the handling of special values. Between FloPoCo cores, this is very advantageous as it saves a lot of decoding and coding blocks for special values. As the proposed architecture contains these blocks, the corresponding FloPoCo conversion blocks InputIEEE and OutputIEEE were used in addition to the FloPoCo SCM block (FPConstMult) to have a fair comparison. The method in [48] is based on adder graphs, so the multiplier block was obtained by the RPAG algorithm without embedded multipliers to ease the comparison. As "round to nearest" rounding instead of the "round to nearest even if tie" [161] scheme is used in the FPConstMult block, this was also selected in the SCM experiment.

A set of single precision constants was chosen as benchmark. Some of the constants are chosen arbitrary while the trigonometric constants appear, e. g., in FFTs [167]. The constants as well as the synthesis results are listed in Table 9.1 (best values are marked bold). It can be observed that the slices could be reduced by 18% while the speed could be improved by 87%

Table 9.1: Benchmark of floating point single constant multiplication

C	FloPoCo [48]				Proposed Method			
	FF	LUT	Slices	f_{max} [MHz]	FF	LUT	Slices	f_{max} [MHz]
$\pi/2$	405	363	130	236.2	**357**	**268**	**102**	**431.4**
$\sqrt{2}$	461	435	154	215.8	**447**	**341**	**134**	**414.4**
e	404	360	133	214.8	**385**	**305**	**110**	**435.0**
$\sin(\pi/3)$	440	422	145	217.3	**374**	**304**	**103**	**429.0**
$\sin(\pi/4)$	461	433	150	237.9	**449**	**337**	**127**	**440.5**
$\sin(\pi/8)$	408	365	137	208.3	**370**	**288**	**101**	**416.5**
$\cos(\pi/8)$	407	360	133	230.4	**357**	**262**	**107**	**429.0**
$\sin(\pi/16)$	**381**	339	117	202.7	384	**267**	**114**	**368.7**
$\cos(\pi/16)$	413	375	140	227.1	**385**	**290**	**117**	**392.6**
$\tan(\pi/16)$	460	370	139	240.5	**416**	**327**	**114**	**411.7**
Average:	424	382	138	223.1	392	299	113	416.9
Improv.:	-	-	-	-	7.5%	21.8%	18.1%	86.9%

on average compared to the FloPoCo cores. The analysis of the VHDL code revealed that this reduction is mainly obtained by more compact adder graphs of the RPAG algorithm. In the FloPoCo block, the adder graphs contain a little more adders (up to two more) and are pipelined afterwards in a heuristic way, leading to additional pipeline registers (up to five more). The latency is similar for both methods. Between 6-7 stages are used in the FloPoCo block while 7 stages are used in the proposed method. Enabling the retiming feature of the synthesis (Xilinx Synthesis Technology (XST) of Xilinx ISE), the speed of FloPoCo could be increased to 359.8 MHz on average. The speed improvement is obtained at the expense of 22 additional slices on average which corresponds to average slice and speed improvements of the proposed method of 28% and 16%, respectively. Hence, always less resources and higher speed could be achieved compared to FloPoCo.

9.3.2 Floating Point MCM

In the second experiment, the single precision floating point MCM circuits of the benchmark filters already used in Section 8.4.2 (see Appendix A.2) were optimized, generated and synthesized. As reference, several floating point SCM cores of FloPoCo [48] and floating point multiplier blocks of Xilinx LogiCORE [168] (obtained from Coregen) were used. The comparison was done for reference designs using slice-only methods (FloPoCo and LogiCORE) and DSP blocks (LogiCORE).

Adder Savings due to Exponents Sharing

The number of adders used in the exponent computation for the conventional method (one adder per output) and the proposed method (sharing exponents) as discussed in Section 9.2.2 are listed in Table 9.2 and visualized in Figure 9.3. Clearly, the sharing of exponents saves resources in all the cases while the reduction is higher for larger filter sizes. While the adder count in the conventional approach grows with a factor of one, it grows with a factor of about one fourth when sharing is used. Of course, filter coefficients are not random and usually do not differ in large magnitudes so exponent sharing may become worse for coefficient values from other applications.

Multiple Constant Multiplication using Slices

For this experiment, RPAG was used without considering embedded multipliers to obtain the multiplier block and LogiCORE [168] was configured to generate logic only and fully pipelined cores. The synthesis results are listed in Table 9.3. The proposed method clearly outperforms the reference methods in all performance metrics. The average slice reductions are 76.9% and 38.7% compared to LogiCORE and FloPoCo, respectively. The corresponding speedups are 15.7% and 61.6% on average. This is no surprise as the reference circuits do not respect any redundancies between different coefficients but it shows the potential of MCM in floating point operations.

Multiple Constant Multiplication using DSP Blocks

In this experiment, the hybrid RPAG approach of Chapter 8 was used to obtain solutions which include DSP blocks. The user defined number of embedded multipliers was set to ∞ to include DSP blocks whenever possible. As reference LogiCORE was configured for pipelined cores with full DSP usage [168]. The synthesis results are listed in Table 9.4. As expected from the mantissa multiplier block results in Chapter 8 (Table 8.3 on page 150), the number of DSP48E1 slices is drastically reduced. Due to the resource sharing a slice reduction of 39% and a speedup of 18% could be achieved on average.

Table 9.2: Required adders for exponents

N_{uq}	Conventional	Shared
5	5	4
10	10	8
20	20	11
30	30	12
40	40	15
50	50	17
60	60	19
70	70	22

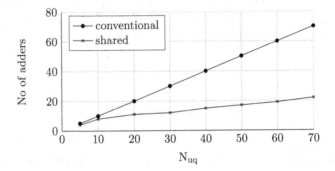

Figure 9.3: Required adders for exponents

Table 9.3: Benchmark of floating point MCM using slices only (best marked bold)

N	Coregen [168] (logic only)		FloPoCo [48]		Proposed Method (logic only)	
	Slices	f_{max}	Slices	f_{max}	Slices	f_{max}
5	1017	289.2	486	210.3	**310**	**333.0**
10	1944	279.9	813	163.4	**554**	**306.5**
20	3843	237.8	1518	167.1	**995**	**300.5**
30	5665	226.5	2214	154.4	**1367**	**233.7**
40	7503	204.3	2681	139.0	**1627**	**255.4**
50	8810	201.2	3271	157.2	**2018**	**228.8**
60	11022	185.0	4191	153.8	**2384**	**218.2**
70	12592	180.0	4531	146.4	**2832**	**211.2**
avg.:	6549.5	225.5	2463.1	161.4	**1510.9**	**260.9**
imp.:	76.93%	15.71%	38.66%	61.61%	-	-

Table 9.4: Benchmark of floating point multiple constant multiplication using logic and embedded DSP48 blocks

N	Coregen (full DSP)			Proposed hybrid Method		
	Slices	DSP48	f_{max}	Slices	DSP48	f_{max}
5	236	10	342.7	**161**	**7**	**349.8**
10	488	20	**289.9**	**301**	**17**	263.9
20	921	40	258.8	**534**	**23**	320.2
30	1329	60	270.9	**827**	**30**	298.2
40	1726	80	248.2	**1035**	**40**	306.8
50	2166	100	222.6	**1352**	**50**	272.4
60	2593	120	214.4	**1593**	**59**	303.8
70	3043	140	203.3	**1803**	**69**	304.3
Avg.:	1562.8	71.25	256.4	**950.8**	**36.9**	**302.4**
Imp.:	-	-	-	39.16%	48.25%	17.97%

9.4 Conclusion

The realization of floating point MCM circuits was discussed and an efficient architecture was proposed in this chapter. Besides the reductions in the integer mantissa multiplier block as introduced in Chapter 8, the sharing of equal exponents and single decoding and sharing of special values can be used to further reduce the complexity of FPMCM circuits.

It was assumed that an IEEE compliant rounding like the "round to nearest, even if tie" scheme is used which requires a full word length result, even if about half of the bits are only computed to decide the rounding direction. A more realistic scenario is the last-bit accurate rounding (also known as faithful [169] or optimal rounding [170]), where the error is defined to be at most one unit in the last place (ulp). A fixed-point faithful rounding architecture for FIR filters with real coefficients was proposed recently by de Dinechin et al. [158].

If the floating point MCM block is used in an FIR filter application, another open issue is an efficient implementation of the remaining structural adders and delays. First experiments showed that the structural adders are often more complex than a single optimized MCM operation. A single pipelined floating point adder with single precision using Xilinx Logi-CORE [168] requires 188 slices (when no DSP block is used). As $N - 1$ structural adders are required for a parallel realization of an N tap FIR filter, the structural adders will dominate the resource usage even for the smallest filter in Table 9.3. Hence, further investigations may be necessary to efficiently merge these floating point adders to avoid repeated normalization steps following the fused datapath concept [171]. One problem to solve is the handling of so-called catastrophic cancellations which may occur in FIR filters and prohibit an efficient faithful rounding [172, 173].

10 Optimization of Adder Graphs with Ternary (3-Input) Adders

Modern FPGAs provide the possibility to realize ternary adders, i. e., adders with three inputs, by using the same hardware resources as for common two-input adders. As shown in Section 2.5.3, the LUTs in RCAs are underutilized or even not used. These LUTs can often not be used to implement other logic functions due to routing constraints. However, the LUTs can be used to build a first 3 : 2 compression stage by using a carry-save adder (CSA) which is then compressed to a single value by the fast carry-chain.

This chapter investigates the use of ternary adders in the PMCM application. The restriction to the PMCM problem is due to the observation of Section 3.3.2 that pipelining is usually necessary on FPGAs to obtain the required performance. For this, the RPAG algorithm of Chapter 4 is extended to support ternary additions. This extension is available in the open-source RPAG implementation (command line flag --ternary_adders) [16]. The motivation to use ternary adders is clearly the reduction of resources. This is illustrated in Figure 10.1, which shows two PAGs for the target coefficient set $T = \{7567, 20406\}$ one using two-input adders and the other using ternary adders. The two sign indicators in the ternary adder nodes correspond to the sign of the second and third input, the first input is positive by definition. It can be observed that the resources can be drastically reduced from nine registered operations to four by using ternary adders. The work described in this chapter was originally published in [174].

10.1 Ternary Adders on Modern FPGAs

A ternary adder performs a summation of the form $S = X + Y + Z$. This can be realized by using a CSA which compresses the three input vectors into two vectors (sum vector and carry vector). Then, these two numbers are added to get the final sum. This principle is shown in Figure 10.2(a).

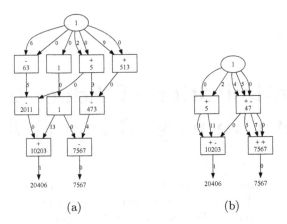

(a)   (b)

Figure 10.1: Pipelined adder graphs of a constant multiplier with coefficients T={7567,20406} using (a) two-input adders (b) ternary adders

The first row of full adders realize the CSA, the second row of full adders corresponds to the RCA. At first view, this saves no hardware resources as the same number of full adders are used compared to two RCA, but on modern FPGAs, the full adders of the CSA can be mapped to the otherwise unused resources of the FPGA LUT. In addition, the first reduction does not comprise any carry-chain which leads to a fast implementation.

10.1.1 Realization on Altera FPGAs

The ALMs, available on Alteras mid-range Arria I,II,V and high-end Stratix II-V FPGAs, contain four 4-input LUTs and two FAs which directly support ternary adders. For that, the ALM supports a special *shared arithmetic mode* in which the output of one LUT is connected to the FA input of the next higher bit [175, 176]. The FAs of the CSA can be directly mapped to the LUTs, realizing sum

$$s'_i = x_i \oplus y_i \oplus z_i \qquad (10.1)$$

and carry

$$c'_i = x_i y_i + x_i z_i + y_i z_i \ . \qquad (10.2)$$

The resulting ternary adder structure is shown in Figure 10.2(b). With each ALM, two output bits can be computed. Note that a two-input adder with

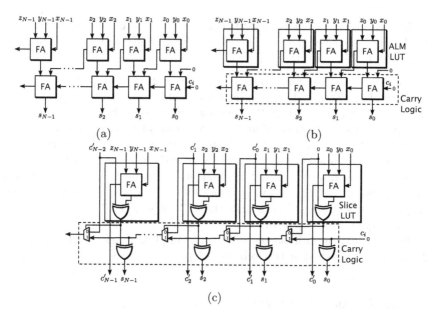

Figure 10.2: Realization of ternary adders on (a) Altera Stratix II-V ALMs (b) Xilinx Virtex 5-7 slices

the same output word size requires the same number of ALMs. The Quartus II tool automatically detects ternary adders from a VHDL statement of the form $s \texttt{ <= } x+y+z$. Unfortunately, the ternary subtract operations where one or more inputs are negated (i.e., $x-y+z$, $x+y-z$ and $x-y-z$) are not directly supported from a high-level description. To overcome this problem, the corresponding input(s) must be inverted and the carry-in signal(s) must be set to 1. To set the carry bit(s), the word size of the adder has to be extended by one bit for $x-y+z$ and $x+y-z$ and by two bits for $x-y-z$, resulting in additional ALMs.

10.1.2 Realization on Xilinx FPGAs

On Xilinx FPGAs, the situation is different as the slice logic does neither provide real full adders nor something comparable to the fast links in the 'shared arithmetic mode'. However, a ternary adder can be efficiently mapped to modern FPGA families which provide six-input LUTs [177], namely the Virtex 5-7, Spartan 6, Kintex 7 and Artix 7 families. In general, one XOR gate has to be realized in the LUT to extend the fast carry logic to a real full adder. In addition to that, the full adder for the CSA is realized in the same LUT. The routing of the carry output to the next higher FA has to be done using the FPGA routing fabric.

The resulting slice configuration is shown in Figure 10.2(c). The multiplexer and the XOR gate shown in the dashed box belong to the carry logic of the slice. The FA and XOR gate shown in the solid boxes are realized in the slice LUTs. With this architecture, four output bits can be computed in each slice. Like for Altera, the same resources are required for a two-input adder with the same output word size. Currently, there is no guarantee that ternary adders from an hardware description language (HDL) description are mapped in that efficient way by the Xilinx tools. Thus, a ternary adder was built by using the Xilinx primitives (`CARRY4` and `LUT6_2`). Note that the ternary subtract operations ($x - y + z$, $x + y - z$ and $x - y - z$) can be realized without additional resources. The implementations for Altera and Xilinx were published at the website of OpenCores [178].

10.1.3 Adder Comparison

As more combinational logic is involved in the critical path of the ternary adder, a speed reduction is expected compared to common two-input adders. Therefore, the speed of ternary adders is experimentally analyzed in this section. For that, adder implementations with two and three synchronized inputs and synchronized outputs were synthesized and analyzed after place & route. This was done for Altera Stratix IV (EP4SGX230KF40C2) using Quartus-II 10.1 and for Xilinx Virtex 6 (XC6VLX75T-2FF484) using ISE 13.4. The resulting clock frequencies for adders with two and three inputs with 16 to 64 bit output word size are shown in Figure 10.3 (top). While both require exactly the same resources the ternary adders are slower. The relative frequency drop from two to three inputs is shown in Figure 10.3 (bottom). While the frequency drop for Stratix IV is only 5% for 16 bit, it is much higher for Virtex 6 (about 40%). Interestingly, the situation for higher word sizes becomes worse on Stratix IV and better on Virtex 6. The large drop for Virtex 6 is caused by the external routing (about 0.5 ns) plus one additional LUT delay (about 0.3 ns) which are large compared to the 15 ps for one bits of carry propagation. For Stratix IV, the internal routing in the 'shared arithmetic mode' is naturally much faster leading to a smaller frequency drop. Note that although the ternary adders for Xilinx use the same slice resources as two-input adders they require more routing resources. This could lead to unusable slice resources in the same region due to routing limitations.

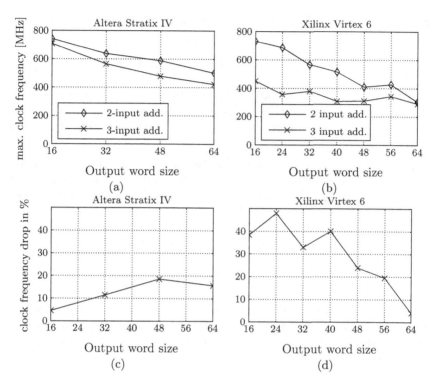

Figure 10.3: Comparison of clock frequencies of two-input and ternary adders (top) and the frequency drop from two to three inputs (bottom) realized on Virtex 6 and Stratix IV FPGAs

10.2 Pipelined MCM with Ternary Adders

In the following section, an extension of the RPAG algorithm of Section 4 to support ternary adders is described. The main extensions are in the definition of $\mathcal{A}$-operation and adder depth as defined for common adders in Section 2.4 and new predecessor topologies which are provided in the following.

10.2.1 $\mathcal{A}$-Operation and Adder Depth

For ternary adders, the $\mathcal{A}$-operation (2.6) is extended to three input arguments:

$$\mathcal{A}_p^3(u,v,w) = |2^{l_u}u + (-1)^{s_v}2^{l_v}v + (-1)^{s_w}2^{l_w}w|2^{-r} \qquad (10.3)$$

The arguments u, v and w correspond to the node inputs and $p = (l_u, l_v, l_w, r, s_v, s_w)$ depicts the configuration, where $l_{u/v/w} \in \mathbb{N}_0$ are left shifts, $r \in \mathbb{N}_0$ is a right shift, and $s_{v/w} \in \{-1, 1\}$ correspond to the signs of v and w.

The same extension can be done for the adder depth (AD). Three partial products can be added in each stage, so the minimum adder depth (AD) for ternary adder graphs is related to the non-zeros (nz) by the base three logarithm

$$\mathrm{AD}_{\min}^3(x) = \lceil \log_3(\mathrm{nz}(x)) \rceil = \left\lceil \frac{\log_2(\mathrm{nz}(x))}{\log_2(3)} \right\rceil \qquad (10.4)$$

$$\leq \lceil 0.631 \cdot \mathrm{AD}_{\min}^2(x) \rceil, \qquad (10.5)$$

where $\mathrm{AD}_{\min}^i(x)$ denotes the minimum adder depth using i-input adders. A comparison of the adder depths (ADs) for adder graphs with two and three inputs is shown in Table 10.1. For $\mathrm{nz}(x) > 4$, using ternary adders always leads to a lower adder depth.

10.2.2 Optimization Heuristic RPAGT

The optimization heuristic is a straight-forward extension of the RPAG algorithm to ternary adders and is called RPAGT in the following. It uses the same overall algorithm as described in Section 4.2.5 and a similar depth search strategy as described in Algorithm 4.3 (page 67) which is shown for reduced pipelined adder graphs with ternary adders (RPAGT) in Algorithm 10.1. Only the best_predecessor_pair is replaced by best_predecessor_set which now computes up to three predecessors and the $\mathcal{A}_*$ set is replaced by its ternary equivalent:

$$\mathcal{A}_*^3(u,v,w) := \{\mathcal{A}_q^3(u,v,w) \mid q \text{ valid configuration}\} \qquad (10.6)$$

$$\mathcal{A}_*^3(X) := \bigcup_{u,v,w \in X} \mathcal{A}_*^3(u,v,w) \qquad (10.7)$$

Table 10.1: Adder depth comparison for two-input and ternary adder graphs

non-zeros nz(x)	1	2	3	4	5-8	9-16	17-26	27-64
$\mathrm{AD}^2_{\min}(x)$	0	1	2	2	3	4	5	5
$\mathrm{AD}^3_{\min}(x)$	0	1	1	2	2	3	3	4

Algorithm 10.1: A single depth search within the RPAGT Algorithm

```
1  RPAGT_DS (X_{s...S}, S)
2    for  s = S...2
3      P := X_{s-1}
4      W := X_s \ A³_*(P)
5      while |W| ≠ 0 do
6        p ← best_single_predecessor(P, W, s)
7        if  p ≠ 0
8          P ← P ∪ {p}
9        else
10         P' ← best_predecessor_set(W, s)
11         P ← P ∪ P'
12       W ← W \ A³_*(P)
13     end while
14     X_{s-1} ← P
```

Like for RPAG with two-input adders, the number of pipeline stages S is set to the minimum possible as it is unlikely that there exists a graph with higher depth but less cost:

$$S := D_{\min} := \max_{t \in T} \mathrm{AD}^3_{\min}(t) \tag{10.8}$$

The main extensions are done in the computation of predecessors which is described in the next section.

10.2.3 Predecessor Topologies

Like in RPAG, a complete search for single predecessors is performed by evaluating all possible topologies (function `best_single_predecessor()` in Algorithm 10.1). These are the topologies (a)-(f) shown in Figure 10.4. Topologies (a)-(c) are identical to topologies (a)-(c) in RPAG (see Figure 4.2 on page 59) as either a register (a) or a two-input adder is used (b)-(c).

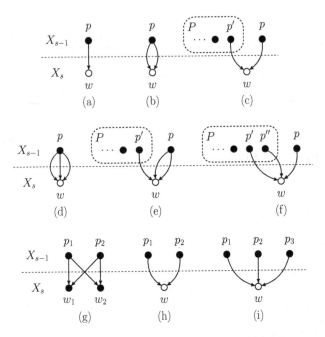

Figure 10.4: Predecessor graph topologies for (a)-(f) one, (g)-(h) two and (i) three predecessors

With ternary adders, there are the three new topologies (d)-(f) for single predecessors. In topology (d), element w is obtained by $p \cdot (2^k \pm 2^l \pm 1)$ with $k, l \in \mathbb{N}_0$. Hence, valid predecessors are all integers obtained by dividing the working set elements by $(2^k \pm 2^l \pm 1)$. If P already contains at least one predecessor, topologies (e) and (f) can be used to compute additional single predecessors. Using Lemma 2 (see page 16), the predecessors in topology (e) are all integers that result from $\mathcal{A}_q(w, p')/(2^k \pm 1)$. It is easy to show that the $\mathcal{A}^3$-operation has the same symmetry properties as stated in Lemma 2 (see page 16). With that we can compute the predecessors of topology (f) by evaluating $\mathcal{A}_q^3(w, p', p'')$. In summary, the computation of all possible single predecessors can be formally represented by the following sets:

$$P_a = \left\{ w \in W \mid \text{AD}^3_{\min}(w) < s \right\} \tag{10.9}$$

$$P_b = \left\{ w/(2^k \pm 1) \mid w \in W,\ k \in \mathbb{N}_0 \right\} \cap \mathbb{N} \setminus \{w\} \tag{10.10}$$

$$P_c = \bigcup_{\substack{w \in W \\ p' \in P}} \left\{ p = \mathcal{A}^2_q(w, p') \mid q \text{ valid},\ \text{AD}^3_{\min}(p) < s \right\} \tag{10.11}$$

$$P_d = \left\{ w/(2^k \pm 2^l \pm 1) \mid w \in W,\ k, l \in \mathbb{N}_0 \right\} \cap \mathbb{N} \setminus P_a \tag{10.12}$$

$$P_e = \bigcup_{\substack{w \in W \\ p' \in P}} \left\{ p = \mathcal{A}^2_q(w, p')/(2^k \pm 1) \mid q \text{ valid}, \dots \right. \tag{10.13}$$
$$\left. \dots \text{AD}^3_{\min}(p) < s,\ k \in \mathbb{N}_0 \right\} \cap \mathbb{N} \setminus P_a$$

$$P_f = \bigcup_{\substack{w \in W \\ p', p'' \in P}} \left\{ p = \mathcal{A}^3_q(w, p', p'') \mid q \text{ valid}, \dots \right. \tag{10.14}$$
$$\left. \dots \text{AD}^3_{\min}(p) < s,\ k \in \mathbb{N}_0 \right\} \setminus P_a$$

These sets are evaluated and the best predecessor out of all the sets is selected.

If no single predecessor was found (indicated by $p = 0$), the search continues with evaluating the topologies (g)-(i). This is done in function best-_predecessor_set() in Algorithm 10.1. Topologies (g)-(h) are identical to the RPAG topologies (d)-(e) as two-input adders are involved. With topology (i), a ternary adder is used to compute w out of three predecessors. As for RPAG, we have to face the problem of too many combinations of valid predecessor triplets. Thus, topology (i) is only evaluated for triplets that result from the MSD representation of the working set elements. Analogous to the two-input case, each predecessor has to have a lower adder depth (AD) and its non-zeros can be computed from

$$\text{nz}(p) \le 3^{\text{AD}^3_{\min}(w) - 1} \tag{10.15}$$

(note the base of 3). Now, all permutations of triplets where each element has less that $\text{nz}(p)$ non-zeros are computed.

To give an example, take the number 53, which has an MSD representation of $53 = 10\bar{1}0101_{\text{MSD}}$. Its MSD representation has four non-zeros which requires at least $S = \lceil \log_3(4) \rceil = 2$ pipeline stages according to (10.5). A valid predecessor can have up to $3^{S-1} = 3$ non-zeros. Therefore, all permutations of triplets having up to three non-zeros are evaluated, which are $P = \{(10\bar{1}, 1, 1), (10001, \bar{1}, 1), (1000001, \bar{1}, 1), (\bar{1}01, 1, 1), (\bar{1}0001, 1, 1), (101, 1, \bar{1}), (10\bar{1}01, 1, 0), (\bar{1}0101, 1, 0)\} = \{(3, 1, 1), (17, -1, 1), (65, -1, 1),$

$(-3, 1, 1)$, $(-15, 1, 1)$, $(5, 1, -1)$, $(13, 1, 0)$, $(-11, 1, 0)\}$. Note that triplets containing a zero can be implemented using ordinary two-input adders.

In principle, topology (g) could be extended to ternary adders. As this would lead to many additional topologies it was not further considered but could be interesting in future work.

10.3 Experimental Results

The RPAGT algorithm was evaluated and compared to the RPAG algorithm for two input adders using the same MIRZAEI10 benchmark as in Chapter 4. The search width limit was set to $L = 100$ like in Section 4.5.6. The results are listed in Table 10.2 wherein the RPAG (v2) results are repeated from Table 4.2 for the sake of comparison. For both algorithms, the number of coefficients N, the number of unique odd coefficients $N_{uq} = |T_{odd}|$, the number of pipeline stages (S), the number of registered adders (reg. add), the number of pure registers (pure reg.) as well as the sum of both registered operations (reg ops) are given.

The use of ternary adders reduces the number of pipeline stages in this benchmark and thereby the latency from $S = 3$ to $S = 2$. The improvement of registered operations of RPAGT over RPAG is shown in the last column of Table 10.2 and is plotted in Figure 10.5. Clearly, the number of registered operations could be reduced for all of the benchmark instances. On average, about 26% less registered operations are required by using ternary adders. It can be observed that the improvement is better for smaller coefficient sets. This can be explained by the fact that the larger filters are getting closer to the lower bound of $N_{uq} + S - 1$ (see Section 4.3) for both architectures. Take, for example, filter instance $N = 151$ which has $N_{uq} = 71$ unique coefficients. This requires 71 adders in the last stage which means that 8 and 3 additional registered operations for non-output fundamentals (NOFs) are used in the solution of RPAG and RPAGT, respectively. This is a great reduction of NOFs but leads to a lower relative improvement in terms of the total number of operations. On average, 1.89 registered operations are needed in addition to the lower bound. This indicates that most instances are either optimal or very close to the optimum. The computation time of RPAGT is slightly larger compared to RPAG but was still less than 30 seconds on a Intel Core i5 CPU.

A VHDL code generator was implemented using the FloPoCo library [2]

Table 10.2: Comparison of high-level operations between RPAG with two-input adders and the ternary adder optimization RPAGT

N	N_{uq}	RPAG (v2)				RPAGT				
		S	reg. add	pure reg.	reg. ops	S	reg. add	pure reg.	reg. ops	impr.
6	3	3	6	3	9	2	5	0	**5**	44.4%
10	5	3	9	3	12	2	6	2	**8**	33.3%
13	7	3	15	0	15	2	8	2	**10**	33.3%
20	10	3	15	3	18	2	9	3	**12**	33.3%
28	14	3	20	3	23	2	15	2	**17**	26.1%
41	21	3	31	1	32	2	23	2	**25**	21.9%
61	31	3	38	3	41	2	32	2	**34**	17.1%
119	54	3	64	3	67	2	56	1	**57**	14.9%
151	71	3	79	4	83	2	72	2	**74**	10.8%
avg.:	24		30.78	2.56	33.33		25.11	1.78	26.89	26.1%

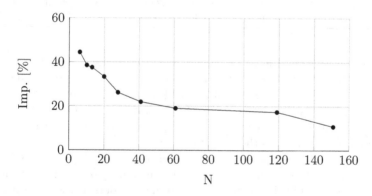

Figure 10.5: Relative improvement of registered operations of RPAGT over RPAG

to realize the optimized PMCM blocks. The VHDL designs were synthesized for Xilinx and Altera FPGAs to evaluate the improvement in terms of logic resources. All synthesis results are listed in Table 10.3. The input word size was chosen as 12 bit as in previous publications [6] and the same synthesis tools and FPGAs as described in Section 10.1.3 were used. The hardware complexity for Altera, measured in adaptive logic modules (ALMs) is reduced by 10.5% on average. As expected, the speed is slightly reduced by 2.7%. For Xilinx, the improvement in terms of slices ranges from 5.8% for the largest coefficient set up to 50% for the smallest instance. On average,

Table 10.3: Synthesis results of MCM blocks optimized by RPAG with two-input adders and the ternary adder optimization RPAGT

	Xilinx Virtex 6				Altera Stratix IV			
	RPAG		RPAGT		RPAG		RPAGT	
N	Slices	f_{max} [MHz]	Slices	f_{max} [MHz]	ALMs	f_{max} [MHz]	ALMs	f_{max} [MHz]
6	50	648.1	25	491.4	108	587.2	62	565.0
10	59	671.1	45	484.3	120	588.2	82	548.0
13	82	630.1	49	441.7	160	592.8	125	546.5
20	87	632.1	65	455.0	183	547.7	141	528.5
28	117	601.3	94	470.2	225	545.9	190	534.8
41	165	598.4	129	434.2	314	540.3	264	522.7
61	217	559.0	190	419.6	434	489.7	397	505.6
119	347	520.0	307	383.6	650	462.3	603	469.0
151	432	482.4	407	374.0	768	445.6	787	450.7
avg.:	172.9	593.6	145.7	439.3	329.1	533.3	294.6	519.0
imp.:	-	-	15.7%	−26.0%	-	-	10.5%	−2.7%

the slice improvement is 15.7%. The maximum frequency is reduced due to the longer critical path of the ternary adders by 26%. However, as the MCM blocks are still very fast, they will most likely not be the limiting factor in a more complex system. With 'more complex', more than two stages of logic in the critical path is meant. The average resource reduction is less than expected if compared to the reduction of registered operations. For Xilinx, this can be explained by the fact that half of the slice flip flops are unused in two-input additions and can be used to implement pure registers without extra resources. This is not the case for ternary adders as the corresponding output is used to compute the carry bit (c_i' in Figure 10.2(c)). For Altera, additional ALUTs are needed in particular for larger MCM instances for negating ternary adder inputs as described in Section 10.1.3.

10.4 Conclusion

The PMCM problem can be extended and solved for ternary adders in a straight forward way as demonstrated by the extension of the RPAG algorithm. In general, more topologies are possible making the optimization slightly more complex but still tractable. The results demonstrate that the use of ternary adders on modern FPGAs allows a great reduction of resources and pipeline stages. Compared to two-input adder PAGs, the number of registered operations (pipelined adders and registers) were reduced by 10% to 44%, depending on the benchmark instance. In terms of logic resources, slice reductions of up to 50% and ALM reductions of up to 42% could be achieved for Xilinx Virtex 6 and Altera Stratix IV FPGAs, respectively. In addition, ternary adder PAGs typically require fewer pipeline stages. The designs were slower compared to two-input adder realizations but still very fast, achieving clock frequencies of typically more than 370 MHz on Virtex 6 and 450 MHz on Stratix IV FPGAs.

11 Conclusion and Future Work

This chapter concludes the thesis by summarizing the obtained results and gives an outlook to future work.

11.1 Conclusion

Different methodologies were investigated in this thesis for the multiple constant multiplication (MCM) computation on FPGAs.

The pipelined MCM (PMCM) optimization methods as presented in the first part of the thesis provide the most universal solution. The optimized pipelined adder graphs (PAGs) can be mapped to nearly any FPGA produced at present and are also applicable to ASICs. Heuristic and optimal algorithms for the PMCM problem were developed and evaluated showing that the direct PMCM optimization is profitable compared to pipelining an already optimized MCM solution obtained by state-of-the-art MCM algorithms. Besides that, it was shown that the MCM problem with bounded or minimal adder depth (AD), which is important for applications requiring low delay and low power, is closely related to the PMCM problem. It can be solved by the proposed PMCM algorithms through little modifications. Furthermore, an extension of the presented heuristic to a related optimization problem, the multiplication of a constant matrix with a vector, was proposed that revealed superior results compared to the state-of-the-art.

The inclusion of FPGA-specific features was investigated in the second part of the thesis. It was shown that considering LUT-based constant multipliers in the PMCM optimization can be beneficial for applications with low input word sizes. For larger input word sizes, a method was developed that efficiently includes a user-defined number of embedded multipliers into the PMCM optimization. It was shown that besides the reduction of adders, MCM techniques can be used to drastically reduce the number of embedded multipliers in large coefficient word size applications, e. g., as needed for floating point operations. An architecture for the MCM operation using floating point arithmetic was proposed in which the sharing of computations is further utilized for an efficient FPGA implementation. Modern FPGAs with 6-input LUTs provide the possibility to realize ternary adders (i. e., three-input adders) with the same resources as common two-input adders. Consequently, a PMCM optimization heuristic for ternary adders was developed. It is the method of choice to realize high efficient logic-based MCM cores.

Overall, different methods were developed to efficiently utilize the capabilities provided by modern FPGAs for MCM. The best method strongly depends on the application requirements (input and coefficient word sizes) and the capabilities of the target device. The proposed methods include solutions for many different requirements and FPGA capabilities. But there are still ideas for future improvements or related computations.

11.2 Future Work

Answering research related questions always leads to new questions (typically more questions arise than have been before). Some issues for future work are addressed in the following, as far as not discussed in the conclusion of the corresponding chapters.

11.2.1 Use of Compressor Trees

It was shown in the last years that the use of generalized parallel counters (GPCs), which are a generalization of full adders, can reduce the critical path of multi-input additions on modern FPGAs [179–184]. It was shown recently, that irregular target optimized GPCs are highly efficient and fast [185]. Furthermore, an ILP based optimization to efficiently use these irregular GPCs in pipelined compressor trees was proposed [186]. Using these techniques, first experiments show that the slice resources of multi-input adders can be halved on average by using GPC compression trees compared to an adder tree using common two-input adders. In addition, the speed was improved by 20%. These advantages could be further exploited for MCM related applications like digital filters. Here, the SOP representation (transposed MCM adder graph) could be realized using GPC compression trees. The main area of interest here is the level of sharing: An optimized SOP circuit with two-input adders would profit from a maximum sharing of intermediate results. Merging all two-input adders into a single compressor tree would allow the use of efficient GPCs but without any sharing. It can be assumed that the optimum is somewhere in between but methods have to be developed to support this to obtain an optimized circuit.

11.2.2 Reconfigurable MCM

An interesting extension to the MCM problem is the optimization of reconfigurable MCM (RMCM) circuits. In an RMCM circuit, the coefficients can be exchanged at runtime, typically with a fixed set of coefficients. Most of the previous work focused on adder graphs which are extended by multiplexers to switch between different coefficients [108, 187–192]. On FPGAs, other resources may be beneficial. Modern Xilinx FPGAs provide so-called configurable LUTs (CFGLUTs), which can be reprogrammed at runtime. The use of CFGLUTs for reconfigurable digital filters was investigated using distributed arithmetic [193] and LUT-based multipliers [194]. The implementations are very efficient and can be programmed with arbitrary coefficients within 32 clock cycles. However, when the configuration time is still too long, the only alternative is the use of multiplexers. Small multiplexers can be fused into the LUT of an RCA [187, 195]. A method to merge several PAGs into a single pipelined RMCM was recently presented [196]. However, instead of merging pre-optimized adder graphs it would be much better to directly optimize the adder graphs such that the merging process results in a solution with even lower cost. One idea to get there is to extend the RPAG algorithm in such a way, that predecessors are not only evaluated for one configuration but for all required configurations. For that, the MUX cost have to be evaluated in the gain function.

11.2.3 Optimal Ternary SCM circuits

Ternary adders can be used on modern FPGAs to reduce the resources in constant multiplication. The largest benefit can be obtained for a low number of coefficients as shown in Chapter 10. Therefore, it would be interesting to search for optimal SCM adder graphs. A similar approach as used in previous work [79, 80] could be performed to enumerate all graph topologies using ternary adders. As pipelining is important on FPGAs, registers should be included in the cost evaluation.

A Benchmark Coefficients

In the following subsections, benchmark coefficients are provided for the sake of reproducibility. Coefficients that are not given here can be found on the FIRSuite website [127].

A.1 Two-Dimensional Filters

The quantized convolution matrices from the 2D filters used in Section 5.6.2 are provided in Table A.1.

Table A.1: Convolution matrices of the benchmark filters

Filter	Convolution Matrix
gaussian 3x3 $B_c = 8\,\text{bit}$	$\begin{pmatrix} 3 & 21 & 3 \\ 21 & 159 & 21 \\ 3 & 21 & 3 \end{pmatrix}$
gaussian 5x5 $B_c = 12\,\text{bit}$	$\begin{pmatrix} 0 & 0 & 1 & 0 & 0 \\ 0 & 46 & 343 & 46 & 0 \\ 1 & 343 & 2534 & 343 & 1 \\ 0 & 46 & 343 & 46 & 0 \\ 0 & 0 & 1 & 0 & 0 \end{pmatrix}$
laplacian 3x3 $B_c = 8\,\text{bit}$	$\begin{pmatrix} 5 & 21 & 5 \\ 21 & -107 & 21 \\ 5 & 21 & 5 \end{pmatrix}$
unsharp 3x3 $B_c = 8\,\text{bit}$	$\begin{pmatrix} -3 & -11 & -3 \\ -11 & 69 & -11 \\ -3 & -11 & -3 \end{pmatrix}$
unsharp 3x3 $B_c = 12\,\text{bit}$	$\begin{pmatrix} -43 & -171 & -43 \\ -171 & 1109 & -171 \\ -43 & -171 & -43 \end{pmatrix}$
lowpass 5x5 $B_c = 8\,\text{bit}$	$\begin{pmatrix} 22 & 88 & 132 & 88 & 22 \\ 88 & 140 & 103 & 140 & 88 \\ 132 & 103 & 106 & 103 & 132 \\ 88 & 140 & 103 & 140 & 88 \\ 22 & 88 & 132 & 88 & 22 \end{pmatrix}$

Filter	Convolution Matrix

lowpass 9x9
$B_c = 10$ bit

```
-1   -7  -25  -50  -62  -50  -25   -7   -1
-7  -25  -10   73  130   73  -10  -25   -7
-25  -10  117  165  126  165  117  -10  -25
-50   73  165  194  303  194  165   73  -50
-62  130  126  303  268  303  126  130  -62
-50   73  165  194  303  194  165   73  -50
-25  -10  117  165  126  165  117  -10  -25
-7  -25  -10   73  130   73  -10  -25   -7
-1   -7  -25  -50  -62  -50  -25   -7   -1
```

lowpass 15x15
$B_c = 12$ bit

```
 0    0    1    5   13   27   40   45   40   27   13    5    1    0    0
 0    2    7   13    2  -41 -103 -133 -103  -41    2   13    7    2    0
 1    7   10  -21  -93 -137 -101  -64 -101 -137  -93  -21   10    7    1
 5   13  -21 -106 -122  -41  -17  -43  -17  -41 -122 -106  -21   13    5
13    2  -93 -122   -8   79  199  304  199   79   -8 -122  -93    2   13
27  -41 -137  -41   79  333  613  662  613  333   79  -41 -137  -41   27
40 -103 -101  -17  199  613  904 1097  904  613  199  -17 -101 -103   40
45 -133  -64  -43  304  662 1097 1197 1097  662  304  -43  -64 -133   45
40 -103 -101  -17  199  613  904 1097  904  613  199  -17 -101 -103   40
27  -41 -137  -41   79  333  613  662  613  333   79  -41 -137  -41   27
13    2  -93 -122   -8   79  199  304  199   79   -8 -122  -93    2   13
 5   13  -21 -106 -122  -41  -17  -43  -17  -41 -122 -106  -21   13    5
 1    7   10  -21  -93 -137 -101  -64 -101 -137  -93  -21   10    7    1
 0    2    7   13    2  -41 -103 -133 -103  -41    2   13    7    2    0
 0    0    1    5   13   27   40   45   40   27   13    5    1    0    0
```

highpass 5x5
$B_c = 8$ bit

```
 -2   -7  -10   -7   -2
 -7   -3    8   -3   -7
-10    8  121    8  -10
 -7   -3    8   -3   -7
 -2   -7  -10   -7   -2
```

highpass 9x9
$B_c = 10$ bit

```
  0   -2   -6  -11  -14  -11   -6   -2    0
 -2   -7  -10   -3    4   -3  -10   -7   -2
 -6  -10   -1   -2  -11   -2   -1  -10   -6
-11   -3   -2  -11   -1  -11   -2   -3  -11
-14    4  -11   -1  500   -1  -11    4  -14
-11   -3   -2  -11   -1  -11   -2   -3  -11
 -6  -10   -1   -2  -11   -2   -1  -10   -6
 -2   -7  -10   -3    4   -3  -10   -7   -2
  0   -2   -6  -11  -14  -11   -6   -2    0
```

highpass 15x15
$B_c = 12$ bit

```
  0    0    0   -1   -3   -6   -8  -10   -8   -6   -3   -1    0    0    0
  0    0   -2   -5   -8   -8   -5   -4   -5   -8   -8   -5   -2    0    0
  0   -2   -6   -9   -8   -6  -10  -13  -10   -6   -8   -9   -6   -2    0
 -1   -5   -9   -8   -8  -14  -14  -11  -14  -14   -8   -8   -9   -5   -1
 -3   -8   -8   -8  -15  -15  -15  -19  -15  -15  -15   -8   -8   -8   -3
 -6   -8   -6  -14  -15  -17  -21  -18  -21  -17  -15  -14   -6   -8   -6
 -8   -5  -10  -14  -15  -21  -19  -23  -19  -21  -15  -14  -10   -5   -8
-10   -4  -13  -11  -19  -18  -23 2028  -23  -18  -19  -11  -13   -4  -10
 -8   -5  -10  -14  -15  -21  -19  -23  -19  -21  -15  -14  -10   -5   -8
 -6   -8   -6  -14  -15  -17  -21  -18  -21  -17  -15  -14   -6   -8   -6
 -3   -8   -8   -8  -15  -15  -15  -19  -15  -15  -15   -8   -8   -8   -3
 -1   -5   -9   -8   -8  -14  -14  -11  -14  -14   -8   -8   -9   -5   -1
  0   -2   -6   -9   -8   -6  -10  -13  -10   -6   -8   -9   -6   -2    0
  0    0   -2   -5   -8   -8   -5   -4   -5   -8   -8   -5   -2    0    0
  0    0    0   -1   -3   -6   -8  -10   -8   -6   -3   -1    0    0    0
```

A.2 Floating-Point Filters

The floating point mantissas used as benchmark coefficients in Chapter 8 are listed in Table A.2 and Table A.3 for single and double precision, respectively.

Table A.2: Mantissas of the single precision benchmark filters

Filter	Mantissa of Floating Point Coefficients
10	9104278 11508221 13820478 13342037 11280171
20	15634560 9224231 8723494 10171660 10944419 13047575 14580070 16152685 14132511 11187164
40	16472687 12086383 8530315 10250822 10968818 10639554 9421781 9651983 10720815 13283749 15358898 9854391 11603900 8792292 8875672 12846475 15279306 8690339 14102557 11152048
60	13969188 14778783 8709645 13144687 13563389 13618183 14581302 9388886 10375529 11389886 12885264 13941351 14574772 14860388 10279381 14227051 16441085 12373066 12694092 13478324 8905724 10835632 12584385 9128746 9292280 12878186 15458187 8829924 14127316 11158130
80	13474641 13747957 15845877 11454604 12942629 10138432 12023284 9707608 11955757 14826523 8846442 11957921 14175421 14627252 9195494 10246435 10707275 11238960 13502040 15114036 16174375 8613996 8818383 9242208 13601877 16611424 9957442 13797790 13995604 14010700 9606441 11463917 13110648 9290230 9554633 12946485 15520137 8916985 14142568 11155790
100	14080712 15982823 11417817 11436200 12839768 15937420 11452001 12971768 13004704 15838718 8896111 10259480 11453045 11607510 13448802 11024101 12110148 9219535 9899282 10457533 14921194 15918209 9052685 10466678 10888896 11146043 13721675 14588600 16493338 8394585 9499605 9689979 10457057 11091633 16080806 9050775 11215968 13838149 14802717 15196098 10119765 11851268 13512535 9427852 9715481 12949570 15596627 8970190 14151872 11159176
120	12209825 14903390 8496278 12644510 8938735 10126789 10252184 10358204 14404497 11130576 12467547 16554096 10172066 14769856 16413738 16639489 16763719 11154424 11858916 15109050 15312420 15567277 8619050 9009482 12635151 8494159 9852003 11230976 14440705 14616179 15737744 9997288 11663061 11726366 12506143 13107472 16267726 8495060 8894124 9516105 10262978 10377976 11532024 12881152 8976488 9679335 12129152 14014517 15461838 16020137 10458149 12160436 13757018 9499179 9841875 12987824 15620059 9011587 14159077 11157325

Filter	Mantissa of Floating Point Coefficients
140	12613137 16114123 16470882 9924283 12445319 9547960 13612845 14466098 12105784 12525141 14223661 14889907 8652066 8790451 9933189 8587407 8834061 16088975 9014581 9699152 13340277 8531322 9865653 10185561 11790636 12243510 14387993 14869981 9383576 10225540 11181106 11824779 12096444 12341700 9269118 10995353 12271215 13464093 16592304 9011359 9306652 11524993 12273357 13465212 13872933 15449107 8908039 9265454 10158570 10536312 10898362 11225604 12200252 14280810 9705575 10077944 12843155 14012565 15883601 16674597 10736527 12359062 13973542 9574115 9923211 12983863 15663812 9038242 14163716 11159699

Table A.3: Mantissas of the double precision benchmark filters

Filter	Mantissa of Floating Point Coefficients
10	4887821980969819 6055995845500609 6178429302891585 7162951821173627 7419812494631982
20	4683390213749697 4952221403961614 5460868568975496 5875740035133365 6006063049480384 7004863398165672 7587334263487904 7827615633786372 8393740459661688 8671906579330022
40	4579677993354586 4665590231493250 4720325796041337 4765090360055590 5058280086484364 5181869181278640 5290535664356144 5503368273665844 5712067058906158 5755693513913230 5888839399348812 5987210008341778 6229796512518356 6488827328663352 6896898712311241 7131658350058977 7571252677905844 8203015149054283 8245745442792976 8843706299120574
60	4675954968324317 4740529544388410 4781224000928108 4900957925304506 4988754736996450 5040619932765449 5518700458222657 5570319667565266 5817335453518010 5990475460725493 6114898523832689 6642739287015346 6756190478629624 6815088872384564 6913923391386445 6917723292546260 7057000030915830 7236120298459812 7281789055161491 7311206266544128 7484705649922286 7499650716399600 7584545252788306 7638090019906398 7824771174529506 7828276649752952 7934298810293632 7978109800215676 8299051067519582 8826740368080341
80	4624603631499059 4734333310914640 4749397317421231 4787269997780586 4936793424763512 4961872772547817 4987654109290647 5129604352277364 5157418797313239 5211732101900242 5345861198975458 5443029103579201 5501012737637710 5748424506880574 5989219358708718 6033870676192532 6149643762494866 6154643789289334 6418698406182950 6419859849993563 6454951637420836 6948520900139372 6950591133546314 7038725586352396 7234142970146972 7248852685199118 7302452010419498 7380878048773080 7407631881046823 7513832714950484 7521937458388016 7592733291023068 7610371408289834 7852945865665576 7959929152913383 8114286284359066 8332310250945998 8507190454094056 8683551487626593 8918190379338350

Filter	Mantissa of Floating Point Coefficients
100	4506808742412963 4776063064814985 4815833881448840 4859097684053159 4860123175407562 4949700031262640 5061539711805544 5100061332956288 5202267712561793 5215959168733095 5314636501968206 5433007456346614 5508016541338521 5614089919557259 5614345310900446 5619254881213138 5845931572810292 5918519197769175 5954774878258850 5983986266233456 5991037131795084 6021527072153706 6129893699665589 6139763035985237 6148245982410772 6148806693376890 6231734733757926 6362601299408845 6501586452706325 6893298169648759 6952247469212815 6964164689097780 6981847070029803 7220270425586419 7254487147777096 7366768129739745 7429299896828553 7559524774758172 7597728476470748 7832195219431610 7947148030394956 8010754806590324 8158342730591069 8373375164308244 8503347075485720 8546023553099620 8556337300181240 8580712730794292 8633317031409708 8854793269618250
120	4560267033048503 4560750664904282 4561404315576820 4627317082406159 4774996494926264 4798946654503876 4819215452149770 4836928667325913 4838058771054530 5099833065268757 5108919710232458 5196553298346095 5283816480746055 5289253797842036 5367252966123845 5436778412674968 5461086167815280 5504099582016434 5509894422720080 5561018637873531 5571633208111139 5614676115240926 5975682616618768 5988485753577772 5990043233777572 6029584405179538 6191207989713223 6261558010334980 6295544909248247 6366706963897266 6511788786217239 6528584629958887 6555099967169407 6693463470672852 6714184607855572 6783445098220591 6788469439265440 6915515622796290 6972784889319138 7037020214883226 7385742550449588 7523986275831315 7601596337944383 7733355397656129 7752794385599424 7847001099343778 7929505974831192 8001196338237613 8111609507090724 8220792809398280 8301010980573662 8357617964795982 8338595053761246 8449136959786580 8600745545855412 8733669022427988 8812058601060438 8887412533948864 8933257901963023 8999953097253862
140	4580218785413783 4610329262639295 4645042753065697 4719337408832784 4742750627222918 4782467159014754 4837936541163180 4839666371195308 4852369288733793 4974352676172526 4976320098210692 4996470990993906 5037769242238784 5126022229784502 5140063679925787 5207192829551992 5210641097764445 5296582100748906 5327483161158829 5328058928974016 5332840256035803 5410554918766731 5453840756450905 5468331397496022 5489794930968042 5656639187794768 5764129214234308 5851013484520190 5903085041779856 5991317676344754 6002810308311294 6026700091989978 6187433476575108 6330049376262800 6348380000115594 6494229036432870 6499243042917251 6549960323880856 6573184375573276 6588058400209046 6589208615119944 6625899904830722 6635220679728058 6681529574467480 6724383620511436 6771626474411076 6895116262637220 6970658534135269 7162006849669238 7228479798315542 7229080863597651 7308340531244791 7447973996360349 7501988232918553 7522938620950963 7604087290459593 7636269785681596 7666951470363230 7724494720254151 7766427002118439 7983260436268202 7993958055398272 8294176290212366 8409445172219169 8527443198785186 8637702431733863 8651204041769903 8842737697683620 8907925601945728 8952106075346954

B Computation of the $\mathcal{A}_*$ Set

The set $\mathcal{A}_*(u, v)$ as defined in Section 2.4.3 (see page 17) seems to be computationally demanding as all combinations of shifts and signs seem to be necessary to evaluate. It is used at several places in the proposed algorithms in this work. As shown next, only a few iterations are actually necessary to compute these sets.

All computations can be limited to odd values of u and v as the even representation can be easily obtained by a bit shift with no cost. In the following, u and v are defined to be odd. The computation of $\mathcal{A}_q(u, v)$ can be divided in three cases:

case $l_u > l_v$:
    In this case, the $\mathcal{A}$-operation (2.6) can be modified to

$$\mathcal{A}_q(u, v) = \left| 2^{l_u - l_v} u + (-1)^{s_v} v \right| 2^{-r + l_v} \tag{B.1}$$

As $2^{l_u - l_v} u$ is even in this case and v is odd per definition, the addition or subtraction of these terms are odd. Therefore, r must be set to $r = l_v$ to fulfill that $\mathcal{A}_q(u, v)$ is odd, too. The computation of the corresponding $\mathcal{A}_*$ elements can be reduced in this case to:

$$\mathcal{A}_*^{(1)}(u, v) = \left\{ \left| 2^k u + (-1)^{s_v} v \right| \mid k = 1 \ldots k_{\max}, \ s_v \in \{0, 1\} \right\} \tag{B.2}$$

case $l_u < l_v$:
    Similar to the last case, the $\mathcal{A}$-operation can be written as

$$\mathcal{A}_q(u, v) = \left| u + (-1)^{s_v} 2^{l_v - l_u} v \right| 2^{-r + l_u} \tag{B.3}$$

resulting in

$$\mathcal{A}_*^{(2)}(u, v) = \left\{ \left| u + (-1)^{s_v} 2^k v \right| \mid k = 1 \ldots k_{\max}, \ s_v \in \{0, 1\} \right\} \tag{B.4}$$

case $l_u = l_v$:
    This is the only case were a right shift is needed to make the result odd:

$$\mathcal{A}_q(u, v) = \left| u + (-1)^{s_v} v \right| 2^{-r + l_u} \tag{B.5}$$

Using the odd-function, it can be written as

$$\mathcal{A}_*^{(3)}(u,v) = \{\text{odd}(|u + (-1)^{s_v} v|) \mid s_v \in \{0,1\}\} \qquad \text{(B.6)}$$

The three cases can be combined to build the complete set:

$$\mathcal{A}_*(u,v) = \mathcal{A}_*^{(1)}(u,v) \cup \mathcal{A}_*^{(2)}(u,v) \cup \mathcal{A}_*^{(3)}(u,v) \qquad \text{(B.7)}$$

The value $k_{\max}$ in (B.2) and (B.4) depends on the actual values of u, v and s_v. The elements in $\mathcal{A}_*(u,v)$ as well as u and v are bounded by $c_{\max}$ (see (2.13) on page 17). The largest shift of k in (B.2) and (B.4) can occur when a small number shifted by k is subtracted by a large number, i.e.,

$$2^k u - v \leq c_{\max} \qquad \text{(B.8)}$$

$$2^k \leq \frac{c_{\max} + v}{u} \leq 2c_{\max} \qquad \text{(B.9)}$$

Hence, shifts of maximally

$$k_{\max} = \log_2(c_{\max}) + 1 = B_T + 2 \qquad \text{(B.10)}$$

have to be evaluated to get all set elements less than $c_{\max}$. Of course, values greater than $c_{\max}$ can occur for $k \leq k_{\max}$ which can be removed after evaluation. Hence, the complexity to compute $\mathcal{A}_*(u,v)$ is linear in relation to the coefficient word size, i.e., $\mathcal{O}(B_T)$.

Bibliography

[1] F. de Dinechin and B. Pasca, "Designing Custom Arithmetic Data Paths with FloPoCo," *IEEE Design & Test of Computers*, vol. 28, no. 4, pp. 18–27, 2012.

[2] F. de Dinechin. (2015, Apr.) FloPoCo Project Website. [Online]. Available: http://flopoco.gforge.inria.fr

[3] M. Wirthlin, "Constant Coefficient Multiplication Using Look-Up Tables," *The Journal of VLSI Signal Processing*, vol. 36, no. 1, pp. 7–15, 2004.

[4] S. Banescu, F. de Dinechin, B. Pasca, and R. Tudoran, "Multipliers for Floating-Point Double Precision and Beyond on FPGAs," *SIGARCH Computer Architecture News*, vol. 38, no. 4, pp. 73–79, Sep. 2010.

[5] S. Mirzaei, R. Kastner, and A. Hosangadi, "Layout Aware Optimization of High Speed Fixed Coefficient FIR Filters for FPGAs," *International Journal of Reconfigurable Computing*, vol. 2010, pp. 1–17, 2010.

[6] S. Mirzaei, A. Hosangadi, and R. Kastner, "FPGA Implementation of High Speed FIR Filters Using Add and Shift Method," in *IEEE International Conference on Computer Design (ICCD)*, 2006, pp. 308–313.

[7] L. Aksoy, E. Costa, P. Flores, and J. Monteiro, "Optimization of Gate-Level Area in High Throughput Multiple Constant Multiplications," in *European Conference on Circuit Theory and Design (ECCTD)*, 2011, pp. 609–612.

[8] M. Kumm, P. Zipf, M. Faust, and C.-H. Chang, "Pipelined Adder Graph Optimization for High Speed Multiple Constant Multiplication," in *IEEE International Symposium on Circuits and Systems (ISCAS)*, 2012, pp. 49–52.

[9] L. Aksoy, E. Gunes, and P. Flores, "An Exact Breadth-First Search Algorithm for the Multiple Constant Multiplications Problem," *NORCHIP*, pp. 41–46, 2008.

[10] M. Cummings and S. Haruyama, "FPGA in the Software Radio,"
     IEEE Communications Magazine, vol. 37, no. 2, pp. 108–112, 1999.

[11] T. Ulversoy, "Software defined radio: Challenges and opportunities,"
     IEEE Communications Surveys & Tutorials, vol. 12, no. 4, pp. 531–
     550, 2010.

[12] M. Potkonjak, M. B. Srivastava, and A. P. Chandrakasan, "Multiple
     Constant Multiplications: Efficient and Versatile Framework and Al-
     gorithms for Exploring Common Subexpression Elimination," *IEEE
     Transactions on Computer-Aided Design of Integrated Circuits and
     Systems*, vol. 15, no. 2, pp. 151–165, 1996.

[13] Y. Wu and J. Li, "The Design of Digital Radar Receivers," *Aerospace
     and Electronic Systems Magazine, IEEE*, vol. 13, no. 1, pp. 35–41,
     1998.

[14] M. M. Appleyard, C. N. Berglund, C. Peterson, and R. W. Smith,
     "Acceleration Management: The Semiconductor Industry Confronts
     the 21st Century," *Technology Management: A Unifying Discipline
     for Melting the Boundaries*, pp. 237–243, 2005.

[15] V. Betz and S. Braun, "FPGA Challenges and Opportunities at
     40nm and Beyond," in *IEEE International Conference on Field Pro-
     grammable Logic and Applications (FPL)*. IEEE, 2009, pp. 4–4.

[16] M. Kumm. (2015, Apr.) Optimization Suite for Pipelined Multiple
     Constant Multiplication. [Online]. Available: http://www.uni-kassel.
     de/go/pagsuite

[17] A. V. Oppenheim and R. W. Schafer, "Discrete-Time Signal Process-
     ing," Prentice Hall Inc., Aug. 2009.

[18] D. Chan and L. Rabiner, "Analysis of Quantization Errors in the Di-
     rect Form for Finite Impulse Response Digital Filters," *IEEE Transac-
     tions on Audio and Electroacoustics*, vol. 21, no. 4, pp. 354–366, Aug.
     1973.

[19] R. E. Crochiere and A. V. Oppenheim, "Analysis of Linear Digital
     Networks," in *Proceedings of the IEEE*, 1975, pp. 581–595.

[20] O. Gustafsson and A. Dempster, "On the Use of Multiple Constant
     Multiplication in Polyphase FIR Filters and Filter Banks," in *Nordic
     Signal Processing Symposium (NORSIG)*, 2004, pp. 53–56.

[21] F. de Dinechin, H. Takeugming, and J.-M. Tanguy, "A 128-Tap Com-
     plex FIR Filter Processing 20 Giga-Samples/S in a Single FPGA," in

Asilomar Conference on Signals, Systems and Computers (ACSSC). IEEE, 2010, pp. 841–844.

[22] W. Han, A. T. Erdogan, T. Arslan, and M. Hasan, "High-Performance Low-Power FFT Cores," *ETRI Journal*, vol. 30, no. 3, pp. 451–460, 2008.

[23] J. L. Shanks, "Computation of the Fast Walsh-Fourier Transform," *IEEE Transactions on Computers*, vol. C-18, no. 5, pp. 457–459, May 1969.

[24] J. Ostermann, J. Bormans, P. List, D. Marpe, M. Narroschke, F. Pereira, T. Stockhammer, and T. Wedi, "Video Coding with H.264/AVC: Tools, Performance, and Complexity," *Circuits and Systems Magazine, IEEE*, vol. 4, no. 1, pp. 7–28, 2004.

[25] International Telecomunication Union (ITU), "Encoding Parameters of Digital Television for Studios," *Recommendation ITU-R BT601-4*, pp. 1–13, 1994.

[26] A. D. Booth, "A Signed Binary Multiplication Technique," *The Quarterly Journal of Mechanics and Applied Mathematics*, pp. 236–240, 1951.

[27] K. D. TOCHER, "Techniques of Multiplication and Division for Automatic Binary Computers," *The Quarterly Journal of Mechanics and Applied Mathematics*, vol. 11, no. 3, pp. 364–384, 1958.

[28] A. Avizienis, "Signed-Digit Number Representations for Fast Parallel Arithmetic," *IEEE Transactions on Electronic Computers*, vol. EC-10, no. 3, pp. 389–400, 1961.

[29] D. E. Knuth, *The Art Of Computer Programming, Volume 2: Seminumerical Algorithms*, 2nd ed. Addison Wesley Publishing Company, 1981.

[30] U. Meyer-Baese, *Digital Signal Processing with Field Programmable Gate Arrays*, 4th ed. Springer Science, 2014.

[31] K. Hwang, *Computer Arithmetic: Principles, Architecture, and Design*. New York: Wiley-Interscience, 1979.

[32] A. V. Aho, S. C. Johnson, and J. D. Ullman, "Code generation for expressions with common subexpressions (Extended Abstract)," in *Proceedings of the rd ACM SIGACT-SIGPLAN Symposium on Principles on Programming Languages POPL*, Jan. 1976.

[33] R. Hartley, "Subexpression Sharing in Filters Using Canonic Signed Digit Multipliers," *IEEE Transactions on Circuits and Systems II: Analog and Digital Signal Processing*, vol. 43, no. 10, pp. 677–688, Oct. 1996.

[34] M. Faust and C.-H. Chang, "Minimal Logic Depth Adder Tree Optimization for Multiple Constant Multiplication," *IEEE International Symposium on Circuits and Systems (ISCAS)*, pp. 457–460, 2010.

[35] O. Gustafsson and H. Johansson, "Efficient Implementation of FIR Filter Based Rational Sampling Rate Converters Using Constant Matrix Multiplication," in *Asilomar Conference on Signals, Systems and Computers (ACSSC)*, 2006, pp. 888–891.

[36] O. Gustafsson, K. Johansson, H. Johansson, and L. Wanhammar, "Implementation of Polyphase Decomposed FIR Filters for Interpolation and Decimation Using Multiple Constant Multiplication Techniques," in *IEEE Asia Pacific Conference on Circuits and Systems (APCCAS)*, 2006, pp. 924–927.

[37] S. Ghissoni, E. Costa, J. Monteiro, and R. Reis, "Combination of Constant Matrix Multiplication and Gate-Level Approaches for Area and Power Efficient Hybrid Radix-2 DIT FFT Realization," in *IEEE International Conference on Electronics, Circuits and Systems, (ICECS)*, 2011, pp. 567–570.

[38] A. Kinane, V. Muresan, and N. O'Connor, "Towards an Optimised VLSI Design Algorithm for the Constant Matrix Multiplication Problem," in *IEEE International Symposium on Circuits and Systems (ISCAS)*, 2006, pp. 5111–5114.

[39] K. Holm and O. Gustafsson, "Low-Complexity and Low-Power Color Space Conversion for Digital Video," in *Norchip Conference*. IEEE, 2006, pp. 179–182.

[40] M. Garrido, F. Qureshi, and O. Gustafsson, "Low-Complexity Multiplierless Constant Rotators Based on Combined Coefficient Selection and Shift-and-Add Implementation (CCSSI)," *IEEE Transactions on Circuits and Systems I: Regular Papers*, pp. 1–11, Jan. 2014.

[41] Y. Voronenko and M. Püschel, "Multiplierless Multiple Constant Multiplication," *ACM Transactions on Algorithms*, vol. 3, no. 2, pp. 1–38, 2007.

[42] A. Dempster and M. D. Macleod, "Constant Integer Multiplication Using Minimum Adders," *IEE Proceedings of Circuits, Devices and Systems*, vol. 141, no. 5, pp. 407–413, 1994.

[43] O. Gustafsson, "A Difference Based Adder Graph Heuristic for Multiple Constant Multiplication Problems," in *IEEE International Symposium on Circuits and Systems (ISCAS)*, 2007, pp. 1097–1100.

[44] A. Dempster and M. Macleod, "Use of Minimum-Adder Multiplier Blocks in FIR Digital Filters," *IEEE Transactions on Circuits and Systems II: Analog and Digital Signal Processing*, vol. 42, no. 9, pp. 569–577, 1995.

[45] K. Johansson, O. Gustafsson, and L. Wanhammar, "A Detailed Complexity Model for Multiple Constant Multiplication and an Algorithm to Minimize the Complexity," *European Conference on Circuit Theory and Design*, vol. 3, pp. III/465–III/468 vol. 3, 2005.

[46] ——, "Bit-Level Optimization of Shift-and-Add Based FIR Filters," in *IEEE International Conference on Electronics, Circuits and Systems, (ICECS)*. IEEE, 2007, pp. 713–716.

[47] L. Aksoy, E. Costa, P. Flores, and J. Monteiro, "Optimization of Area in Digital FIR Filters using Gate-Level Metrics," in *ACM/IEEE Design Automation Conference (DAC)*, 2007, pp. 420–423.

[48] N. Brisebarre, F. de Dinechin, and J.-M. Muller, "Integer and Floating-Point Constant Multipliers for FPGAs," *IEEE International Conference on Application-Specific Systems, Architectures and Processors (ASAP)*, pp. 239–244, 2008.

[49] H.-J. Kang and I.-C. Park, "FIR Filter Synthesis Algorithms for Minimizing the Delay and the Number of Adders," *IEEE Transactions on Circuits and Systems II: Analog and Digital Signal Processing*, vol. 48, no. 8, pp. 770–777, 2001.

[50] S. Demirsoy, A. Dempster, and I. Kale, "Transition Analysis on FPGA for Multiplier-Block Based FIR Filter Structures," *IEEE International Symposium on Circuits and Systems (ISCAS)*, vol. 2, pp. 862–865 vol.2, 2000.

[51] A. G. Dempster, S. S. Demirsoy, and I. Kale, "Designing Multiplier Blocks with Low Logic Depth," in *IEEE International Symposium on Circuits and Systems (ISCAS)*. IEEE, 2002, pp. V–773–V–776.

[52] S. Demirsoy, A. Dempster, and I. Kale, "Power Analysis of Multiplier Blocks," in *IEEE International Symposium on Circuits and Systems (ISCAS)*, 2002, pp. I–297–I–300 vol.1.

[53] K. Johansson, "Low Power and Low Complexity Shift-and-Add Based Computations," Ph.D. dissertation, Linköping Studies in Science and Technology, 2008.

[54] K. Johansson, O. Gustafsson, and L. S. DeBrunner, "Minimum Adder Depth Multiple Constant Multiplication Algorithm for Low Power FIR Filters," in *IEEE International Symposium on Circuits and Systems (ISCAS)*, 2011, pp. 1439–1442.

[55] D. Shi and Y. J. Yu, "Design of Linear Phase FIR Filters With High Probability of Achieving Minimum Number of Adders," *IEEE Transactions on Circuits and Systems I: Regular Papers*, vol. 58, no. 1, pp. 126–136, 2011.

[56] S. D. Brown, R. J. Francis, J. Rose, and Z. G. Vranesic, *Field-Programmable Gate Arrays*. Kluwer Academic Publishers, 2014.

[57] V. Betz, J. Rose, and A. Marquardt, *Architecture and CAD for Deep-Submicron FPGAS*. Kluwer Academic Publishers, 1999.

[58] B. Pasca, "High-Performance Floating-Point Computing on Reconfigurable Circuits," Ph.D. dissertation, École Normale Supérieure de Lyon, 2012.

[59] Xilinx, Inc., *Virtex-4 FPGA User Guide (UG070)*, Dec. 2008.

[60] ——, *Virtex-5 FPGA User Guide (UG190)*, Dec. 2007.

[61] Xilinx Inc., *Virtex-6 FPGA Configurable Logic Block User Guide (UG364)*, Feb. 2012.

[62] Xilinx, Inc., *Spartan-6 FPGA Configurable Logic Block User Guide (UG384)*, Feb. 2010.

[63] ——, *7 Series FPGAs Configurable Logic Block Users Guide (UG474)*, Nov. 2012.

[64] Altera Corporation, *Stratix III Device Handbook, Volume 1*, Mar. 2011.

[65] ——, *Stratix IV Device Handbook Volume 1*, Sep. 2012.

[66] ——, *Stratix V Device Handbook*, Jun. 2013.

[67] F. de Dinechin, H. Nguyen, and B. Pasca, "Pipelined FPGA Adders," in *IEEE International Conference on Field Programmable Logic and Application (FPL)*, 2010, pp. 422–427.

[68] Xilinx, Inc., *Virtex-6 FPGA Data Sheet: DC and Switching Characteristics (DS152)*, Mar. 2014.

[69] M. J. Flynn and S. F. Oberman, *Advanced Computer Arithmetic*. New York: John Wiley & Sons Inc., 2001.

[70] B. Parhami, *Computer Arithmetic - Algorithms and Hardware Designs*. Oxford University Press, Oct. 2009.

[71] S. Xing and W. W. H. Yu, "FPGA Adders: Performance Evaluation and Optimal Design," *IEEE Design & Test of Computers*, vol. 15, no. 1, pp. 24–29, 1998.

[72] M. Rogawski, E. Homsirikamol, and K. Gaj, "A Novel Modular Adder for One Thousand Bits and More Using Fast Carry Chains of Modern FPGAs," *IEEE International Conference on Field Programmable Logic and Application (FPL)*, pp. 1–8, 2014.

[73] K. K. Parhi, *VLSI Digital Signal Processing Systems: Design and Implementation*. John Wiley & Sons, 1999.

[74] O. Gustafsson, "Lower Bounds for Constant Multiplication Problems," *IEEE Transactions on Circuits and Systems II: Express Briefs*, vol. 54, no. 11, pp. 974–978, Nov. 2007.

[75] Xilinx Inc., *Virtex-6 Libraries Guide for HDL Designs*, Jun. 2011.

[76] C. B. Boyer and U. C. Merzbach, *A History of Mathematics*. John Wiley & Sons, Jan. 2011.

[77] J. Colson, "A Short Account of Negativo-Affirmative Arithmetick," *Philosophical Transactions*, vol. 34, 1726.

[78] D. R. Bull and D. H. Horrocks, "Primitive Operator Digital Filters," *IEE Proceedings of Circuits, Devices and Systems*, vol. 138, no. 3, pp. 401–412, Jun. 1991.

[79] O. Gustafsson, A. Dempster, and L. Wanhammar, "Extended Results for Minimum-Adder Constant Integer Multipliers," in *IEEE International Symposium on Circuits and Systems (ISCAS)*, 2002, pp. 73–76.

[80] O. Gustafsson, A. Dempster, K. Johansson, M. Macleod, and L. Wanhammar, "Simplified Design of Constant Coefficient Multipliers," *Cir-

cuits, Systems, and Signal Processing, vol. 25, no. 2, pp. 225–251, 2006.

[81] J. Thong and N. Nicolici, "A Novel Optimal Single Constant Multiplication Algorithm," in *ACM/IEEE Design Automation Conference (DAC)*, Jun. 2010.

[82] ——, "An Optimal and Practical Approach to Single Constant Multiplication," *IEEE Transactions on Computer-Aided Design of Integrated Circuits and Systems*, vol. 30, no. 9, pp. 1373–1386, Sep. 2011.

[83] P. Cappello and K. Steiglitz, "Some Complexity Issues in Digital Signal Processing," *IEEE Transactions on Acoustics, Speech and Signal Processing*, vol. 32, no. 5, pp. 1037–1041, Oct. 1984.

[84] N. Sankarayya, K. Roy, and D. Bhattacharya, "Algorithms for Low Power and High Speed FIR Filter Realization Using Differential Coefficients," *IEEE Transactions on Circuits and Systems II: Analog and Digital Signal Processing*, vol. 44, no. 6, pp. 488–497, Jun. 1997.

[85] R. Pasko, P. Schaumont, V. Derudder, S. Vernalde, and D. Durackova, "A New Algorithm for Elimination of Common Subexpressions," *IEEE Transactions on Computer-Aided Design of Integrated Circuits and Systems*, vol. 18, no. 1, pp. 58–68, 1999.

[86] I.-C. Park and H.-J. Kang, "Digital Filter Synthesis Based on Minimal Signed Digit Representation," in *Design, Automation and Test in Europe (DATE)*, 2001, pp. 468–473.

[87] C.-Y. Yao, H.-H. Chen, T.-F. Lin, C.-J. Chien, and C.-T. Hsu, "A Novel Common-Subexpression-Elimination Method for Synthesizing Fixed-Point FIR Filters," *IEEE Transactions on Circuits and Systems I: Regular Papers*, vol. 51, no. 11, pp. 2215–2221, 2004.

[88] O. Gustafsson and L. Wanhammar, "A Novel Approach to Multiple Constant Multiplication Using Minimum Spanning Trees," *IEEE Midwest Symposium on Circuits and Systems (MWSCAS)*, vol. 3, 2002.

[89] A. Vinod, E.-K. Lai, A. Premkuntar, and C. Lau, "FIR Filter Implementation by Efficient Sharing of Horizontal and Vertical Common Subexpressions," *Electronics Letters*, vol. 39, no. 2, pp. 251–253, 2003.

[90] O. Gustafsson, "Contributions to Low-Complexity Digital Filters," Ph.D. dissertation, Linköping Studies in Science and Technology, Department of Electrical Engineering, Linköping University, 2003.

[91] O. Gustafsson, H. Ohlsson, and L. Wanhammar, "Improved Multiple Constant Multiplication Using a Minimum Spanning Tree," in *Asilomar Conference on Signals, Systems and Computers (ACSSC)*, 2004, pp. 63–66.

[92] N. Boullis and A. Tisserand, "Some Optimizations of Hardware Multiplication by Constant Matrices," in *IEEE Transactions on Computers*, Oct. 2005, pp. 1271–1282.

[93] F. Xu, C.-H. Chang, and C.-C. Jong, "Contention Resolution Algorithm for Common Subexpression Elimination in Digital Filter Design," *IEEE Transactions on Circuits and Systems II: Express Briefs*, vol. 52, no. 10, pp. 695–700, 2005.

[94] P. Flores, J. Monteiro, and E. Costa, "An Exact Algorithm for the Maximal Sharing of Partial Terms in Multiple Constant Multiplications," in *IEEE/ACM International Conference on Computer Aided Design (ICCAD)*, 2005, pp. 13–16.

[95] L. Aksoy, E. Costa, P. Flores, and J. Monteiro, "Optimization of Area Under a Delay Constraint in Digital Filter Synthesis Using SAT-Based Integer Linear Programming," *ACM/IEEE Design Automation Conference (DAC)*, pp. 669–674, 2006.

[96] L. Aksoy, E. da Costa, P. Flores, and J. Monteiro, "Exact and Approximate Algorithms for the Optimization of Area and Delay in Multiple Constant Multiplications," *IEEE Transactions on Computer-Aided Design of Integrated Circuits and Systems*, vol. 27, no. 6, pp. 1013–1026, 2008.

[97] O. Gustafsson, "Towards Optimal Multiple Constant Multiplication: A Hypergraph Approach," in *Asilomar Conference on Signals, Systems and Computers (ACSSC)*. IEEE, Oct. 2008, pp. 1805–1809.

[98] L. Aksoy, "Optimization Algorithms for the Multiple Constant Multiplications Problem," Ph.D. dissertation, Istanbul Technical University, 2009.

[99] L. Aksoy, E. Günes, and P. Flores, "Search Algorithms for the Multiple Constant Multiplications Problem: Exact and Approximate," *Microprocessors and Microsystems*, vol. 34, no. 5, pp. 151–162, 2010.

[100] J. Thong and N. Nicolici, "Combined Optimal and Heuristic Approaches for Multiple Constant Multiplication," in *IEEE International Conference on Computer Design (ICCD)*. IEEE, 2010, pp. 266–273.

[101] L. Aksoy, E. Costa, P. Flores, and J. Monteiro, "Finding the Optimal Tradeoff Between Area and Delay in Multiple Constant Multiplications," *Microprocessors and Microsystems*, 2011.

[102] J. Thong and N. Nicolici, "Time-Efficient Single Constant Multiplication Based on Overlapping Digit Patterns," *IEEE Transactions on Very Large Scale Integration Systems (VLSI)*, vol. 17, no. 9, Sep. 2009.

[103] A. Dempster, M. D. Macleod, and O. Gustafsson, "Comparison of Graphical and Subexpression Methods for Design of Efficient Multipliers," in *Asilomar Conference on Signals, Systems and Computers (ACSSC)*, 2004, pp. 72–76.

[104] Spiral Project Website. (2015, Apr.) Software/Hardware Generation for DSP Algorithms. [Online]. Available: http://www.spiral.net

[105] H. J. Prömel and A. Steger, *The Steiner Tree Problem*, ser. Advanced Lectures in Mathematics. Wiesbaden: Vieweg+Teubner Verlag, 2002.

[106] D. Romero and U. Meyer-Baese, "On the Inclusion of Prime Factors to Calculate the Theoretical Lower Bounds in Multiplierless Single Constant Multiplications," *EURASIP Journal on Advances in Signal Processing*, 2014.

[107] K. Johansson, L. DeBrunner, O. Gustafsson, and V. DeBrunner, "Design of Multiplierless FIR Filters with an Adder Depth Versus Filter Order Trade-Off," in *Asilomar Conference on Signals, Systems and Computers (ACSSC)*, 2009, pp. 744–748.

[108] M. Faust, O. Gustafsson, and C.-H. Chang, "Reconfigurable Multiple Constant Multiplication Using Minimum Adder Depth," in *Asilomar Conference on Signals, Systems and Computers (ACSSC)*, 2010, pp. 1297–1301.

[109] L. Aksoy, E. Costa, P. Flores, and J. Monteiro, "Design of Low-Power Multiple Constant Multiplications Using Low-Complexity Minimum Depth Operations," in *Proceedings of the Great Lakes Symposium on VLSI (GLSVLSI)*. ACM, 2011, pp. 79–84.

[110] X. Lou, Y. J. Yu, and P. K. Meher, "High-Speed Multiplier Block Design Based on Bit-Level Critical Path Optimization," *IEEE International Symposium on Circuits and Systems (ISCAS)*, pp. 1308–1311, 2014.

[111] ——, "Fine-Grained Critical Path Analysis and Optimization for Area-Time Efficient Realization of Multiple Constant Multiplications,"

IEEE Transactions on Circuits and Systems I: Regular Papers, vol. 62, no. 3, pp. 863–872, Jul. 2015.

[112] O. Gustafsson, H. Ohlsson, and L. Wanhammar, "Minimum-Adder Integer Multipliers Using Carry-Save Adders," *IEEE International Symposium on Circuits and Systems (ISCAS)*, vol. 2, pp. 709–712, 2001.

[113] O. Gustafsson, A. Dempster, and L. Wanhammar, "Multiplier Blocks Using Carry-Save Adders," in *IEEE International Symposium on Circuits and Systems (ISCAS)*, 2004, pp. 473–476.

[114] O. Gustafsson, H. Ohlsson, and L. Wanhammar, "Carry-Save Adder Based Difference Methods for Multiple Constant Multiplication in High-Speed FIR Filters," in *National Conf. Radio Science (RVK)*, *Linköping, Sweden*, 2005, pp. 245–248.

[115] O. Gustafsson and L. Wanhammar, "Low-Complexity and High-Speed Constant Multiplications for Digital Filters Using Carry-Save Arithmetic," in *Digital Filters.* InTech, Apr. 2011.

[116] ——, "Low-Complexity Constant Multiplication Using Carry-Save Arithmetic for High-Speed Digital Filters," *International Symposium on Image and Signal Processing and Analysis (ISPA)*, pp. 212–217, 2007.

[117] A. Blad and O. Gustafsson, "Integer Linear Programming-Based Bit-Level Optimization for High-Speed FIR Decimation Filter Architectures," *Circuits, Systems, and Signal Processing*, vol. 29, no. 1, pp. 81–101, Jun. 2009.

[118] K. Macpherson and R. Stewart, "Area Efficient FIR Filters for High Speed FPGA Implementation," *IEE Proceedings on Vision, Image and Signal Processing*, vol. 153, no. 6, pp. 711–720, Dec. 2006.

[119] U. Meyer-Baese, J. Chen, C. H. Chang, and A. G. Dempster, "A Comparison of Pipelined RAG-n and DA FPGA-based Multiplierless Filters," *IEEE Asia Pacific Conference on Circuits and Systems (APC-CAS)*, pp. 1555–1558, Dec. 2006.

[120] S. Mirzaei, "Design Methodologies and Architectures for Digital Signal Processing on FPGAs," Ph.D. dissertation, University of California Santa Barbara, 2010.

[121] L. Aksoy, E. Costa, P. Flores, and J. Monteiro, "Optimization of Area and Delay at Gate-Level in Multiple Constant Multiplications," in *Eu-*

romicro Conference on Digital System Design: Architectures, Methods and Tools. IEEE, 2010, pp. 3–10.

[122] A. P. Chandrakasan, S. Sheng, and R. W. Brodersen, "Low-power CMOS digital design," *IEEE Journal of Solid-State Circuits*, vol. 27, no. 4, pp. 473–484, 1992.

[123] A. Correale Jr, "Overview of the Power Minimization Techniques Employed in the IBM PowerPC 4xx Embedded Controllers," in *International Symposium on Low Power Design ISLPED*, Apr. 1995.

[124] S. J. Wilton, S.-S. Ang, and W. Luk, "The Impact of Pipelining on Energy per Operation in Field-Programmable Gate Arrays," in *IEEE International Conference on Field Programmable Logic and Applications (FPL)*. Springer, 2004, pp. 719–728.

[125] M. Kumm and P. Zipf, "High Speed Low Complexity FPGA-Based FIR Filters Using Pipelined Adder Graphs," in *IEEE International Conference on Field-Programmable Technology (FPT)*, 2011, pp. 1–4.

[126] Mathworks, Inc., *Matlab External Interfaces*, 2015.

[127] FIRsuite. (2015, Apr.) Suite of Constant Coefficient FIR Filters. [Online]. Available: http://www.firsuite.net

[128] T. Achterberg, "Constraint Integer Programming," Ph.D. dissertation, Technische Universität Berlin, 2007.

[129] Konrad-Zuse-Zentrum für Informationstechnik Berlin. (2015, Apr.) A MIP Solver and Constraint Integer Programming Framework. [Online]. Available: http://scip.zib.de

[130] Xilinx, Inc., *FIR Compiler v4.0 (DS534)*, Jun. 2008.

[131] University of California, San Diego. (2015, Apr.) Website of the Department of Computer Science and Engineering. [Online]. Available: http://kastner.ucsd.edu/fir-benchmarks/

[132] M. J. Beauchamp, S. Hauck, K. D. Underwood, and K. S. Hemmert, "Architectural Modifications to Enhance the Floating-Point Performance of FPGAs," *IEEE Transactions on Very Large Scale Integration Systems (VLSI)*, vol. 16, no. 2, pp. 177–187, 2008.

[133] Xilinx, Inc., *Virtex-II Platform FPGAs: Complete Data Sheet (DS031)*, Nov. 2007.

[134] ——, *XtremeDSP for Virtex-4 FPGAs User Guide (UG073)*, May 2008.

[135] R. L. Graham, D. E. Knuth, and O. Patashnik, *Concrete Mathematics - A Foundation for Computer Science*. Addison-Wesley Professional, Jan. 1994.

[136] Free Software Foundation, Inc. (2015, May) GCC, the GNU Compiler Collection. [Online]. Available: https://gcc.gnu.org

[137] University of Illinois. (2015, May) clang: A C Language Family Frontend for LLVM. [Online]. Available: http://clang.llvm.org

[138] Graphviz. (2015, May) Graphviz - Graph Visualization Software. [Online]. Available: http://graphviz.org/

[139] M. Kumm, D. Fanghänel, K. Möller, P. Zipf, and U. Meyer-Baese, "FIR Filter Optimization for Video Processing on FPGAs," *EURASIP Journal on Advances in Signal Processing (Springer)*, pp. 1–18, 2013.

[140] S. Voß, "The Steiner Tree Problem with Hop Constraints," *Annals of Operations Research*, vol. 86, pp. 321–345, 1999.

[141] C. E. Miller, A. W. Tucker, and R. A. Zemlin, "Integer Programming Formulation of Traveling Salesman Problems," *Journal of the ACM*, vol. 7, no. 4, pp. 326–329, Oct. 1960.

[142] R. Hassin and D. Segev, "The Set Cover with Pairs Problem," *Foundations of Software Technology and Theoretical Computer Science (FSTTCS)*, pp. 164–176, 2005.

[143] L. B. Gonçalves, S. de Lima Martins, L. S. Ochi, and A. Subramanian, "Exact and Heuristic Approaches for the Set Cover with Pairs Problem," *Optimization Letters*, vol. 6, no. 4, pp. 641–653, 2011.

[144] Google Inc. (2015, Apr.) Google Optimization Tools. [Online]. Available: https://developers.google.com/optimization/

[145] IBM. (2015, Apr.) High-Performance Mathematical Programming Solver for Linear Programming, Mixed Integer Programming, and Quadratic Programming. [Online]. Available: http://www.ibm.com/software/commerce/optimization/cplex-optimizer/

[146] Gurobi Optimization Inc. (2015, Apr.) Gurobi Website. [Online]. Available: http://www.gurobi.com

[147] C. Lavin, M. Padilla, J. Lamprecht, P. Lundrigan, B. Nelson, and B. Hutchings, "RapidSmith: Do-It-Yourself CAD Tools for Xilinx FPGAs," in *IEEE International Conference on Field Programmable Logic and Application (FPL)*, 2011.

[148] R. Hartley, "Optimization of Canonic Signed Digit Multipliers for Filter Design," in *IEEE International Symposium on Circuits and Systems (ISCAS)*, 1991, pp. 1992–1995.

[149] A. Dempster, O. Gustafsson, and J. O. Coleman, "Towards an Algorithm for Matrix Multiplier Blocks," in *European Conference on Circuit Theory and Design (ECCTD)*, 2003, pp. 1–4.

[150] O. Gustafsson, H. Ohlsson, and L. Wanhammar, "Low-Complexity Constant Coefficient Matrix Multiplication Using a Minimum Spanning Tree Approach," in *Proceedings of the 6th Nordic Signal Processing Symposium (NORSIG)*, 2004, pp. 141–144.

[151] M. D. Macleod and A. Dempster, "Common Subexpression Elimination Algorithm for Low-Cost Multiplierless Implementation of Matrix Multipliers," *Electronics Letters*, vol. 40, no. 11, pp. 651–652, 2004.

[152] A. Kinane, V. Muresan, and N. O'Connor, "Optimisation of Constant Matrix Multiplication Operation Hardware Using a Genetic Algorithm," in *Lecture Notes in Computer Science*. Springer, 2006, pp. 296–307.

[153] U. Meyer-Baese, G. Botella, D. Romero, and M. Kumm, "Optimization of High Speed Pipelining in FPGA-Based FIR Filter Design Using Genetic Algorithm," in *Proceedings of SPIE, the International Society for Optical Engineering*, 2012, pp. 1–12.

[154] H. S. Malvar, A. Hallapuro, M. Karczewicz, and L. Kerofsky, "Low-Complexity Transform and Quantization in H.264/AVC," *IEEE Transactions on Circuits and Systems for Video Technology*, vol. 13, no. 7, pp. 598–603, 2003.

[155] K. D. Chapman, "Fast Integer Multipliers Fit in FPGAs," *Electronic Design News*, 1994.

[156] K. Chapman, "Constant Coefficient Multipliers for the XC4000E," *Xilinx Application Note*, 1996.

[157] M. Faust and C.-H. Chang, "Bit-parallel Multiple Constant Multiplication using Look-Up Tables on FPGA," *IEEE International Symposium on Circuits and Systems (ISCAS)*, pp. 657–660, 2011.

[158] F. de Dinechin, M. Istoan, and A. Massouri, "Sum-of-Product Architectures Computing Just Right," *IEEE International Conference on Application-Specific Systems, Architectures and Processors (ASAP)*, pp. 41–47, 2014.

[159] F. de Dinechin, "Multiplication by Rational Constants," *accepted for publication in IEEE Transactions on Circuits and Systems II: Express Briefs*, no. 99, pp. 1–5, 2012.

[160] M. Kumm and P. Zipf, "Hybrid Multiple Constant Multiplication for FPGAs," in *IEEE International Conference on Electronics, Circuits and Systems, (ICECS)*, 2012, pp. 556–559.

[161] "IEEE Standard for Floating-Point Arithmetic," 2008.

[162] A. Karatsuba and Y. Ofman, "Multiplication of Multidigit Numbers on Automata," *Soviet Physics Doklady*, vol. 7, p. 595, Jan. 1963.

[163] F. de Dinechin and B. Pasca, "Large Multipliers with Fewer DSP Blocks," in *IEEE International Conference on Field Programmable Logic and Application (FPL)*, 2009, pp. 250–255.

[164] L. Aksoy, P. Flores, and J. Monteiro, "Efficient design of FIR filters using hybrid multiple constant multiplications on FPGA," in *IEEE International Conference on Computer Design (ICCD)*. IEEE, 2014, pp. 42–47.

[165] M. Kumm, K. Liebisch, and P. Zipf, "Reduced Complexity Single and Multiple Constant Multiplication in Floating Point Precision," in *IEEE International Conference on Field Programmable Logic and Application (FPL)*, 2012, pp. 255–261.

[166] "IEEE Standard for Binary Floating-Point Arithmetic," 1985.

[167] F. Qureshi and O. Gustafsson, "Low-Complexity Reconfigurable Complex Constant Multiplication for FFTs," in *IEEE International Symposium on Circuits and Systems (ISCAS)*, 2009, pp. 1137–1140.

[168] Xilinx Inc., *LogiCORE IP Floating-Point Operator v6.0*, 2012.

[169] T. J. Dekker, "A Floating-Point Technique for Extending the Available Precision," *Numerische Mathematik*, vol. 18, no. 3, pp. 224–242, Jun. 1971.

[170] U. Kulisch, "An Axiomatic Approach to Rounded Computations," *Numerische Mathematik*, vol. 18, no. 1, pp. 1–17, Feb. 1971.

[171] M. Langhammer, "Floating Point Datapath Synthesis for FPGAs," in *IEEE International Conference on Field Programmable Logic and Application (FPL)*, 2008, pp. 355–360.

[172] M. Daumas and D. W. Matula, "Design of a Fast Validated Dot Product Operation," in *IEEE Symposium on Computer Arithmetic (ARITH)*, 1993, pp. 62–69.

[173] B. Lee and N. Burgess, "Parameterisable Floating-Point Operations on FPGA," in *Asilomar Conference on Signals, Systems and Computers (ACSSC)*, 2002, pp. 1064–1068.

[174] M. Kumm, M. Hardieck, J. Willkomm, P. Zipf, and U. Meyer-Baese, "Multiple Constant Multiplication with Ternary Adders," in *IEEE International Conference on Field Programmable Logic and Application (FPL)*, 2013, pp. 1–8.

[175] G. Baeckler, M. Langhammer, J. Schleicher, and R. Yuan, "Logic Cell Supporting Addition of Three Binary Words," *US Patent No 7565388, Altera Coop.*, 2009.

[176] Altera Corporation, *Stratix II Device Handbook*, 2005.

[177] J. M. Simkins and B. D. Philofsky, "Structures and Methods for Implementing Ternary Adders/Subtractors in Programmable Logic Devices," *US Patent No 7274211, Xilinx Inc.*, Sep. 2007.

[178] OpenCores. (2013) Ternary Adder IP Core. [Online]. Available: http://opencores.org/project,ternary_adder

[179] H. Parandeh-Afshar, P. Brisk, and P. Ienne, "Efficient Synthesis of Compressor Trees on FPGAs," in *Asia and South Pacific Design Automation Conference (ASPDAC)*. IEEE, 2008, pp. 138–143.

[180] ——, "Improving Synthesis of Compressor Trees on FPGAs via Integer Linear Programming," in *Design, Automation and Test in Europe (DATE)*. IEEE, 2008, pp. 1256–1261.

[181] H. Parandeh-Afshar, A. Neogy, P. Brisk, and P. Ienne, "Compressor Tree Synthesis on Commercial High-Performance FPGAs," *ACM Transactions on Reconfigurable Technology and Systems (TRETS)*, vol. 4, no. 4, pp. 1–19, 2011.

[182] T. Matsunaga, S. Kimura, and Y. Matsunaga, "Power and Delay Aware Synthesis of Multi-Operand Adders Targeting LUT-Based FPGAs," *International Symposium on Low Power Electronics and Design (ISLPED)*, pp. 217–222, 2011.

[183] ——, "Multi-Operand Adder Synthesis Targeting FPGAs," *IEICE Transactions on Fundamentals of Electronics, Communications and Computer Sciences*, vol. E94-A, no. 12, pp. 2579–2586, Dec. 2011.

[184] ——, "An Exact Approach for GPC-Based Compressor Tree Synthesis," *IEICE Transactions on Fundamentals of Electronics, Communications and Computer Sciences*, vol. E96-A, no. 12, pp. 2553–2560, Dec. 2013.

[185] M. Kumm and P. Zipf, "Efficient High Speed Compression Trees on Xilinx FPGAs," in *Methoden und Beschreibungssprachen zur Modellierung und Verifikation von Schaltungen und Systemen (MBMV)*, 2014.

[186] ——, "Pipelined Compressor Tree Optimization Using Integer Linear Programming," in *IEEE International Conference on Field Programmable Logic and Application (FPL)*. IEEE, 2014, pp. 1–8.

[187] S. S. Demirsoy, I. Kale, and A. Dempster, "Reconfigurable Multiplier Blocks: Structures, Algorithm and Applications," *Circuits, Systems, and Signal Processing*, vol. 26, no. 6, pp. 793–827, 2007.

[188] P. Tummeltshammer, J. C. Hoe, and M. Püschel, "Time-Multiplexed Multiple-Constant Multiplication," *IEEE Transactions on Computer-Aided Design of Integrated Circuits and Systems*, vol. 26, no. 9, pp. 1551–1563, Sep. 2007.

[189] J. Chen and C.-H. Chang, "High-Level Synthesis Algorithm for the Design of Reconfigurable Constant Multiplier," *IEEE Transactions on Computer-Aided Design of Integrated Circuits and Systems*, vol. 28, no. 12, pp. 1844–1856, Dec. 2009.

[190] L. Aksoy, P. Flores, and J. Monteiro, "Towards the Least Complex Time-Multiplexed Constant Multiplication," in *IFIP/IEEE 21st International Conference on Very Large Scale Integration (VLSI-SoC)*, 2013.

[191] ——, "Optimization of Design Complexity in Time-Multiplexed Constant Multiplications," in *Design, Automation & Test in Europe Conference & Exhibition (DATE)*, 2014, pp. 1–4.

[192] ——, "Multiplierless Design of Folded DSP Blocks," *ACM Transactions on Design Automation of Electronic Systems (TODAES)*, vol. 20, no. 1, pp. 14–24, Nov. 2014.

[193] M. Kumm, K. Möller, and P. Zipf, "Reconfigurable FIR Filter Using Distributed Arithmetic on FPGAs," in *IEEE International Symposium on Circuits and Systems (ISCAS)*, 2013, pp. 2058–2061.

[194] ——, "Dynamically Reconfigurable FIR Filter Architectures with Fast Reconfiguration," *International Workshop on Reconfigurable Communication-centric Systems-on-Chip (ReCoSoC)*, pp. 1–8, 2013.

[195] K. Möller, M. Kumm, B. Barschtipan, and P. Zipf, "Dynamically Reconfigurable Constant Multiplication on FPGAs." in *Methoden und Beschreibungssprachen zur Modellierung und Verifikation von Schaltungen und Systemen (MBMV)*, 2014, pp. 159–169.

[196] K. Möller, M. Kumm, M. Kleinlein, and P. Zipf, "Pipelined Reconfigurable Multiplication with Constants on FPGAs," in *IEEE International Conference on Field Programmable Logic and Application (FPL)*. IEEE, 2014, pp. 1–6.

Printed in the United States
By Bookmasters